Safety Design Criteria for Industrial Plants

Volume II

Editors

Maurizio Cumo
Antonio Naviglio
Professors
University of Rome
Rome, Italy

CRC Press
Taylor & Francis Group
Boca Raton London New York

CRC Press is an imprint of the
Taylor & Francis Group, an **informa** business

CRC Press
Taylor & Francis Group
6000 Broken Sound Parkway NW, Suite 300
Boca Raton, FL 33487-2742

Reissued 2019 by CRC Press

© 1989 by Taylor & Francis Group, LLC
CRC Press is an imprint of Taylor & Francis Group, an Informa business

No claim to original U.S. Government works

A Library of Congress record exists under LC control number:

Publisher's Note
The publisher has gone to great lengths to ensure the quality of this reprint but points out that some imperfections in the original copies may be apparent.

Disclaimer
The publisher has made every effort to trace copyright holders and welcomes correspondence from those they have been unable to contact.

ISBN 13: 978-0-367-25974-7 (hbk)
ISBN 13: 978-0-367-25976-1 (pbk)
ISBN 13: 978-0-429-29086-2 (ebk)

Visit the Taylor & Francis Web site at http://www.taylorandfrancis.com and the
CRC Press Web site at http://www.crcpress.com

DEDICATION

To my son Fabrizio

Maurizio Cumo

To my parents Anna Maria and Luigi Eros

Antonio Naviglio

PREFACE

The aim of this book is to provide a picture of safety-related aspects of the design, operation, and maintenance of a high-risk plant or — more generally — of a plant where substances potentially harmful for human health are handled.

The book is addressed to people looking for an introduction to safety design criteria of industrial plants, namely students, engineers, and technicians involved in the design, operation, or maintenance of industrial plants, people working for local or national authorities or agencies, with a responsibility in licensing or control of industrial plants.

The book is not an exhaustive handbook including all the aspects and methodologies employed for the design of industrial plants using hazardous matter, but it is an introductory guide addressing the reader to the most relevant items and aspects that should be known when operating, designing, and licensing this kind of plant.

The most relevant concepts and methodologies are introduced and a continuous reference to specialized books or papers dealing with the specific themes is done to allow a deeper insight for people looking for a better knowledge.

A trial has been done for a unique treatment for all kinds of medium and high-risk plants; explicit references to the specific kinds of plants are used whenever necessary.

The wide range of topics, including general safety design criteria, risk analysis methodologies, the main damage causes for the plants, the effects of dispersion of pollutants into the environment, and the human health effects of most relevant pollutants is aimed at giving a picture of the various aspects that could or should be considered in the safety or risk assessment of an industrial plant.

The description of the various external and internal events that could prove as initiating events of accidents does not imply that all of them must be considered in the design of a plant; what is important is to know their existence, while the applicability of the single design criterion will be judged by the designer, also on the basis of the rules and regulations locally in force.

The subject has been subdivided into six sections.

The first, dedicated to the concept and definition of industrial risks, is aimed at introducing the concept of the technological risk (Chapter 1) and at the quantitative indentification of the risk for human health, relative to the main high-risk substances (chemical, radiological) (Chapter 2).

Section II, dedicated to the main aspects of risk analyses of industrial plants, includes the definitions of the main magnitudes commonly employed in risk analyses and assessment (Chapter 3), the main phases of such analyses and assessment (Chapter 4), and the description of methodologies for a system reliability assessment (Chapter 5).

The third section, dedicated to the main design criteria of a general validity and to the causes of damage for the components of an industrial plant, internal to the process, includes a description of general safety design criteria (Chapter 6), the analysis of causes of damage potentially responsible for a release of hazardous substances of thermodynamic nature (Chapter 7) and chemical nature (Chapter 8), and the analysis of the relevance and of the role of instrumentation and control for the operation and protection of risky plants and for the timely monitoring of hazardous matters releases (Chapter 9).

The fourth section assumes that damage to a plant might happen and analyzes the dispersion mechanisms of pollutants into the environment, with the aim of identifying methods to limit the exposure to risk for population and operation personnel. It includes a description of methods for the evaluation of the consequences of effluents dispersion (Chapter 10), methods to limit exposure during normal operation of plants (Chapter 11), and methods to limit exposure following accidents (Chapter 12).

Section I in Volume II, dedicated to the potential causes of damage for the components of an industrial plant external to the process, includes the analysis and criteria to defend the plant against natural events, comprehending the soil-structure interactions (Chapter 1), flooding (Chapter 2), the analysis and design criteria against man-induced events, comprehending fire (Chapter 3), clouds explosions, toxic clouds, and sabotage actions (Chapter 4), and area events (Chapter 5); an insight is performed on the role of human factors for the safety of plants (Chapter 6).

The first five sections are especially addressed to plants handling high-risk substances and to associated criteria to limit the entity of the health risk; Section II in Volume II, on the contrary, is focused on "conventional risks" in industrial plants, which are by far mainly responsible for injuries to personnel involved in operation and maintenance of industrial plants. This section includes an analysis of general industrial risk-protection criteria and a description of most common health risks for personnel operating in industrial plants.

The relevance of the analyzed topics, together with their multidisciplinarity, would justify a very deep analysis; they are only introduced in this book, whose aim is to give a general feeling on the variety and complexity of aspects to be considered, so as to allow the most appropriate decisions in view of further research. Nevertheless, the modern approach to safety is of a global nature, considering and comparing the different points of view to build a system assessment which needs specific contributions of different nature and sciences in harmonic construction. In this sense the book may offer a method and a sensibility for the new mind which is necessary to master the safety of high-risk installations.

Maurizio Cumo
Antonio Naviglio
University of Rome
May 31, 1987

THE EDITORS

Maurizio Cumo is Professor of Nuclear Plants at the University La Sapienza of Rome, where he is also director of the postgraduate School for Nuclear Safety and Radioprotection.

He received his doctorate in Nuclear Engineering at the Politechnic University of Milan in 1962. Since that year, without discontinuities, he has participated in much thermohydraulic research at the ENEA Research Center Casaccia near Rome as well as at the University of Rome. In this field he has authored or coauthored more than 150 scientific publications both international and national as well as 2 books. A member of the Assembly for International Heat Transfer Conferences, the Executive Committee of the International Center for Heat and Mass Transfer, the EUROTHERM Committee, and the European Two-Phase Flow Group with particular attention to experimental developments, presently he is acting as President of the Italian Commission for Nuclear Safety and Health Protection and as a member of the board of the Italian State Agency for Nuclear and Alternative Energies (ENEA).

He is also Chairman of the Italian Association of Nuclear Engineering (ANDIN), Vice Chairman of the Italian Society of Standards (UNI), a member of the board of directors of the International Solar Energy Society (ISES), the American Nuclear Society, the American Institute of Chemical Engineers, and the New York Academy of Sciences.

Biographical references are provided by Who's Who in the World, Who's Who in Europe, International Book of Honor, International Who's Who in Education, Dictionary of International Biography, World Nuclear/World Energy Directory, and the International Directory of Distinguished Leadership.

Antonio Naviglio is Professor of Thermal Hydraulics in the Department of Energetics, Faculty of Engineering, University of Rome.

He received the Italian "Laurea" in Nuclear Engineering in 1973 from the University of Rome. During 1973 to 1975 he worked as a process engineer in a major Italian engineering company. In 1975 and 1976 he worked for the Italian Agency for Nuclear Safety and Radiological Protection (ENEA-DISP). During 1976 to 1981 he worked as a process engineer mainly in the field of thermal hydraulics, for the Italian Electric Power Authority (ENEL). Since 1981, he has been working at the University of Rome, first as assistant professor and then as Professor of Thermal Hydraulics.

Professor Naviglio is an expert member of the Italian Committee for Nuclear Safety and Radiological Protection, the executive committee of ANDIN, the Italian Association of Nuclear Engineers and of ANIAI, the Italian Association of Architects and Engineers, and the Director of UNITAR/UNDP Centre on Small Energy Resources in Rome.

Professor Naviglio has authored or coauthored some 90 scientific publications in the field of heat transfer and energy exploitation.

The research activity of Professor Naviglio has been mainly devoted to heat transfer phenomena, to complex thermal hydraulic phenomenologies affecting the performance and safety of equipment both for nuclear and for chemical plants, and to the development of innovative processes allowing energy saving and minimizing environmental impact.

CONTRIBUTORS

Claudia Bartolomei
Facoltà Di Ingegneria
Università Degli Studi Di Roma
Rome, Italy

Alberto Ferreli
ENEA
Rome, Italy

Leonardo Lojelo
TecnoIdroMeteo
Rome, Italy

Francesco Mazzini
Direttore del
 Centro Studi ed Esperienze
Corpo Nazionale Vigili del Fuoco
Rome, Italy

Rinaldo Paciucci
IAEIR
Rome, Italy

Sergio Paribelli
ENEA Sede Centrale
Rome, Italy

Salvatore Ragusa
Managing Director
S.I. Safety Improvement
Milan, Italy

Ivo Tripputi
ENEL - DCO
Rome, Italy

TABLE OF CONTENTS

Volume I

Volume II

Section I
External Damage Causes

INTRODUCTION

In Section I and Section IV of Volume I the main safety criteria to be considered by the designer of a plant handling hazardous substances have been analyzed.

Such criteria are aimed at avoiding damages to the plant, thus avoiding release of the hazardous substances.

The safety philosophy cannot be limited, nevertheless, to the prevention aspect; the relevance of consequences of releases of high-risk substances imposes the consideration of accidents that — in spite of the preventive measures adopted in the design — might however occur.

In addition, several processes where noxious substances are handled are characterized by releases not limited to accidental conditions but during normal operation (continuous or batch). It is, therefore, important to be able to evaluate the modes of dispersion of effluents into the environment, so as to be able to evaluate the exposure for humans, plants, and animals. The knowledge of the physics of dispersion of harmful substances allows the identification of the consequences of the normal or accidental releases; it is the basic point for the study of actions able to limit or reduce the consequences of releases that in any case cannot be reduced, so as for the identification of maximum values of releases acceptable during normal operation or to hypothesize as consequences of accidents (and to which to bind the design itself), if limits on the doses (more generally, on the consequences) are fixed.

Chapter 1

NATURAL EVENTS: SEISMIC AND GEOTECHNICAL ASPECTS

Leonardo Lojelo

TABLE OF CONTENTS

I. INTRODUCTION

This chapter deals with interactions between the plant and the underlying soil region. In particular, two main aspects are accounted for: the seismic aspect and the geotechnical aspect. Both are particularly important for site selection and plant design and both may represent a hazard for the operation of the plant. The two aspects, in some cases, are strictly correlated. For this reason and for exposition purposes, all the geotechnical issues related to the earthquake-resistant design of the plant will be dealt, in the following, as seismic aspect.

II. THE SEISMIC ASPECT: EARTHQUAKE-RESISTANT DESIGN

In the last 10 years a great deal of information and records about earthquakes have been collected and analyzed allowing a rapid progress in the state of the art of seismology and earthquake engineering. Even if, today, many questions about the phenomenon and its effects need a complete answer, it is possible to realize how the seismic energy is released at the source and evaluate how the seismic waves propagate. Moreover, engineers manage to estimate the effects of the ground motion on the structures and are able to design them in an earthquake-resistant way. The knowledge about earthquakes increased as industrialization and technical progress increased. In fact, new seismic design and construction requirements have been developed when new structural systems, materials, and construction techniques replaced the traditional ones.

The design of industrial facilities, whose failure could have catastrophic consequences, such as nuclear power plants, stressed the necessity to protect them against earthquakes. It should be emphasized, in fact, that a great impulse to new studies and research in the field of seismology and earthquake engineering is due to the increased safety request about nuclear power plants and at this request is now involving the other major hazard industrial plants.

The criteria and methodologies, developed for the seismic protection of nuclear facilities, can be used as a guide for the design of major hazard industrial plants together with the national building codes for seismic design which represent in any case a minimum requirement.

A. Earthquake Characteristics

Earthquakes may have different origins.[1] Most of them have a tectonic origin: they result from the sudden release of the strain energy, which accumulates within the outer shell of the earth along discontinuities or fractures of the crustal rock (the faults).

Plate tectonic theory[2] explains the mechanism of the strain energy accumulation. According to this theory, the lithosphere is made up by plates that continuously change and move independently from the others (Figure 1). The plates slide away from the mid-ocean ridges, where dense magma from earth interior creates new matter, while the old matter dives back where an oceanic plate collides with the edge of a continental plate, sinking below it. At the edges of the plates, where they collide or move along in opposite direction, tectonic forces strain the crustal rock. If the strength of the rock is overcome, then the rock breaks, coming back to its unstrained shape, and the elastic rebound generates seismic waves.[3] The failure takes place on the weak point of the crust, generally where a discontinuity of the rock (a fault) already exists, and propagates from it. The point, where is the first release of energy, is called ''hypocenter'' and its projection at the ground surface is called ''epicenter''.

The seismic waves that propagate from the hypocenter are body waves: longitudinal and shear waves. The first ones (also called P or primary waves) travel faster and produce a compression-dilatation motion in the direction of propagation (Figure 2A). The shear waves (also known as S or secondary waves) travel slower but, generally, with greater amplitude and periods. They generate a distortional motion, with no volume variation, at a right angle with the propagation direction (Figure 2B).

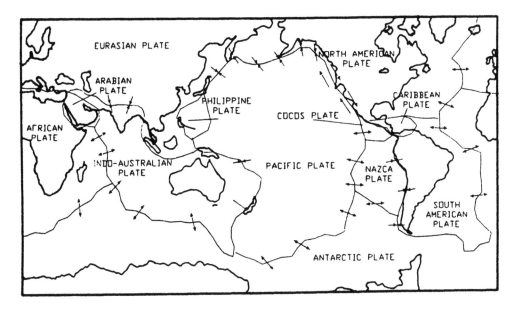

FIGURE 1. Configuration and relative movements of the lithospheric plates.

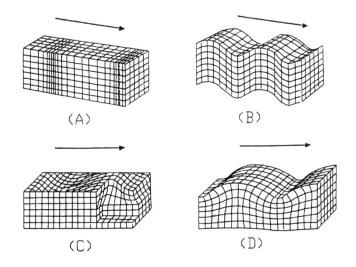

FIGURE 2. Seismic waves: (A) longitudinal waves, (B) shear waves, (C) Love waves, and (D) Rayleigh waves.

P and S waves, arriving at the ground surface, create surface waves: Love (L) and Rayleigh (R) waves, which propagate along the surface only. A transversal motion on a horizontal plane is associated to the L-waves (Figure 2C), while the R-waves generate an elliptic motion on a vertical plane in the propagation direction (Figure 2D).

Traveling towards the ground surface the seismic waves change their characteristics (amplitude and frequency content) because of reflection, refraction, and attenuation phenomena.

The upper soil layers, generally alluvial deposits, finally work as a filter amplifying or deamplifying the motion, depending on the frequency range.

All the above-mentioned phenomena, that can be summarized as (Figure 3):

- Source mechanism

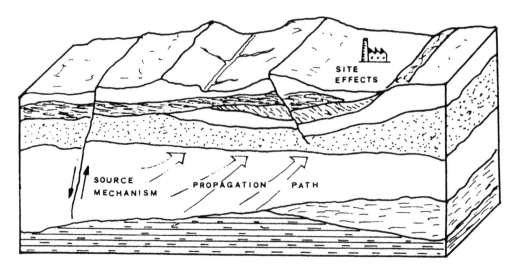

FIGURE 3.   Phenomena which influence the seismic ground motion at a specific site.

- Seismic wave propagation path
- Site amplification and deamplification effects,

strongly influence the characteristics of the seismic motion, experienced at a point on the ground surface, which are

- Peak ground acceleration, velocity, and displacement
- Frequency content
- Duration

Of course, the influence of these phenomena on the ground motion is different for different situations. For example, far from the epicenter, on an alluvial deposit, the propagation path and the site amplification-deamplification effects are very important, while, near the epicenter, on a rock site, the ground motion characteristics are particularly affected by the source mechanism.

1. Source Mechanism

All the physical processes, which take place at the hypocenter and which generate the energy release as seismic waves, are identified as "source mechanism".

The physics of these processes are still not completely understood. It is believed that the characteristics of the generated seismic waves depend on many parameters, such as the fault type (normal, thrust, or strike-slip), fault displacement, rupture area, rupture velocity, type of rupture (single or multiple), stress drop, elastic modulus of the rock at the source, and source depth.

Information about the source mechanism of an earthquake can be derived by elaborating the seismographic records. In particular it is possible to compute the "magnitude" M of the earthquake: a number which is a characteristic of the energy released at the source. The local magnitude (M_l), first introduced by Richter[4] in 1935, is defined as the logarithm of the maximum amplitude, expressed in micron, recorded by a standard seismograph (a Wood-Anderson apparatus with specified constants) situated on a firm ground at an epicentral distance of 100 km. Other scales of magnitude are available: the body-wave magnitude (M_b) and the surface-wave magnitude (M_s). They are defined, respectively, considering the amplitude of the seismic waves with 1- and 20-s period.

Table 1
PEAK GROUND MOTION ATTENUATION RELATIONSHIPS

Attenuation relationship	Applicability	Ref.
$a = 0.018 \exp (0.868\ M)\ [D + 0.0606 \exp(0.700\ M)]^{-1.09}(1 + 0.45p)$	Worldwide	8
$\log a = -1.02 + 0.249\ M - \log[(D^2 + 7.3^2)^{0.5}] - 0.00255(D^2 + 7.3^2)^{0.5} + 0.26p$	Western north America	9
$\log v = -0.67 + 0.489\ M - \log[(D^2 + 4.0^2)^{0.5}] - 0.00256(D^2 + 4.0^2)^{0.5} + 0.17s + 0.22p$ (s = 0: rock site and s = 1: soil site)		
$\log a = -1.562 + 0.306\ M - \log[(D^2 + 5.8^2)^{0.5}] + 0.169s + 0.173p$ (s = 0: stiff and deep soil site and s = 1: shallow soil site)	Italy	10
$\log v = -0.710 + 0.455\ M - \log[(D^2 + 3.6^2)^{0.5}] + 0.133s + 0.215p$ (s = 0: stiff soil site and s = 1 shallow and deep soil site)		

Note: a: peak horizontal ground acceleration (g); v: peak horizontal ground velocity (cm/s); M: magnitude; D: closest distance to the source (km); and p = 0 mean value, and p = 1 mean-plus-SD value.

Another way to characterize an earthquake is represented by the intensity: a numerical index, expressed in Roman numerals, which describes the effects of earthquakes on earth surface, on man, and on buildings. The intensity scales are particularly useful to characterize the historical earthquakes, for which there are no instrumental records. Unlike the magnitude, the intensity is a qualitative index and presents some interpretative problems due to the population density and the type of buildings, existing in the region at the age of the earthquake. The intensity at the epicenter and the felt area, defined by the contours with equal intensity, can be empirically correlated to the magnitude.[5,6]

2. Seismic Wave Propagation Path

One of the most important aspects about the seismicity of a region is the attenuation of the seismic motion with the distance from the source. This phenomenon depends on anisotropy and inhomogeneity of the medium crossed by the seismic waves, on reflection, refraction, canalization, scattering, wave-mode conversion, absorption, and wave interference.

Regional ground motion attenuation relationships have been developed from strong motion records, obtained at different distances from the source, during the same earthquake or during earthquakes with similar magnitude. These relationships allow to evaluate the mean and the mean-plus-a-standard-deviation value of the peak ground acceleration and, in some cases, velocity as a function of the earthquake magnitude and the distance from the source. A comprehensive summary of selected strong motion attenuation relationships, proposed from 1974 to 1984, has been presented by Campbell.[7] Table 1 presents an example of attenuation relationships: the first one[8] has been obtained using worldwide data, while the other two[9,10] have been determined from regional data (North American and Italian data). Finally, Table 2 presents relationships[11,12] between the peak ground motion parameters (acceleration, velocity, and displacement) derived for different site conditions.

The information about the distribution of the seismic intensity in a region provides also an alternative way, even if qualitative, of evaluating the seismic wave propagation path. In fact the isoseismic contours (felt area) allow to evaluate the attenuation of the seimic intensity with the distance from the epicenter.

3. Site Amplification and Deamplification Effects

It is widely recognized that the stratigraphic characteristics and the physical and mechanical properties of the geologic materials, underlying a particular site, influence the ground motion at the surface, modifying, in many cases significantly, the amplitude and the frequency content. It has been documented, in fact, that, during earthquakes, the building damage, in

Table 2
SUGGESTED RELATIONSHIPS
BETWEEN PEAK GROUND
ACCELERATION, VELOCITY,
AND DISPLACEMENT

Relationship	Ref.
Soil: $\dfrac{v}{a}$ = 121.9 cm/sec/g	11
Rock: $\dfrac{v}{a}$ = 91.4 cm/sec/g	
$\dfrac{ad}{v^2}$ = 6	
Stiff soil: $\dfrac{v}{a}$ = 110 cm/sec/g	12
Rock: $\dfrac{v}{a}$ = 55 cm/sec/g	
Deep[a] stiff soil: $\dfrac{v}{a}$ = 135 cm/sec/g	

Note: a: peak horizontal ground acceleration;
v: peak horizontal ground velocity; and
d: peak horizontal ground displacement.

[a]   depth > 60 m.

the same area, is a function of the foundation soil characteristics;[13,14] greater damage is likely to occur when the natural frequencies of the buildings are similar to that of the soil deposits on which they are constructed.

A schematic geologic situation that allows to explain, in a simple way, site amplification-deamplification effects is presented in Figure 4A.

Site S is on a soil deposit overlying a bedrock, which outcrops at site R. Assuming that sites S and R are at the same distance from the source, the differences between the ground motions, recorded at the surface of the two sites during the same earthquake, are only due to soil deposit effects. The ratio, in the frequency domain, between the ground motions recorded at sites S and R (transfer function) shows that the soil deposit amplifies the rock outcropping motion in well-defined frequency ranges, while deamplification effects appear in the other frequency ranges (Figure 4B).

The peak values of the transfer function correspond to the natural frequencies of the soil deposit. They gradually lower as the frequency increases as an effect of the soil material damping. With the hypothesis of vertically propagating shear waves, the natural frequency of the soil deposit, corresponding to the nth mode of vibration, is given by:[15]

$$f = (2n - 1)\frac{V_s}{4H} \tag{1}$$

where V_s is the shear wave velocity in the soil and H is the soil deposit thickness.

The corresponding peak values of the transfer function, can be determined, for small values of the material damping β_s, by the following:[15]

$$\frac{U_s}{U_R} = \frac{1}{\dfrac{V_s \, \rho_s}{V_R \, \rho_R} + \beta_s(2n - 1)\dfrac{\pi}{2}} \tag{2}$$

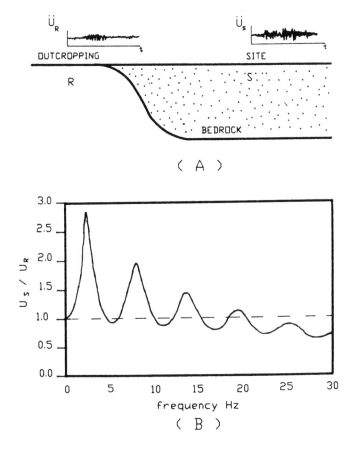

FIGURE 4. Site amplification-deamplification effects: (A) schematic geologic situation, and (B) transfer function.

Where ρ_s and ρ_R are the mass density of the soil and the rock and V_R is the shear wave velocity in the rock.

For real cases the site amplification-deamplification phenomenon depends on many parameters: shear wave velocity and propagation direction, soil density, soil material damping and nonlinear behavior, slope and thickness of layers, geometry of lenses, and ground surface topography. In most cases (level ground deposits with horizontal layers), the site amplification-deamplification effects can be evaluated by the theory of vertically propagating shear waves through a viscoelastic layered medium.[16]

B. Earthquake-Resistant Design Criteria

Earthquake-resistant design must take into account all the effects due to earthquakes that can cause damage and represent a threat to human lives. The following aspects shall be considered specifically for major hazard industrial plants:

- Collapse of the buildings
- Damage to structural elements
- Damage to nonstructural elements
- Damage to systems and components
- Damage due to nearby structure interaction
- Damage due to deformations and failure of the foundation soil
- Damage due to landslides
- Indirect consequences such as floods, fire, and explosions

The above-mentioned effects could take place at the same time, superimposing, because earthquakes contemporary affect every part of the buildings and all that is inside. Moreover earthquakes may cause damage to different parts of the plant, located in different buildings on the same site.

If there is a concentration of industrial facilities in the same area, an accident to a plant may cause accidents to the other plants, which are in a vulnerable situation, having suffered the effects of the same earthquake. Therefore, in site selection it may be important to avoid not only highly seismic areas or hazards due to foundation soil stability but concentration of major hazard industrial facilities as well.

The earthquake-resistant design of the structures, systems, and components of a plant take into consideration the following steps:

- Definition of the design earthquake
- Modeling of soil, structure, systems, and components
- Evaluation of the seismic response
- Stability of foundation soil

The above-mentioned points qualify the design of the plant, as a decision taken in one point can heavily condition the subsequent choices. Therefore, it is essential that the operated choices and the applied safety margins are consistent to the level of seismic protection that the social community wants to apply.

It should be noted that different earthquake-resistant design requirements can be applied according to the potential hazard of the plant. Moreover, inside the same plant, not all the buildings, systems and components have necessarily to be designed according to the same requirements. More severe seismic requirements are applied to the structures, systems, and components of the plant, which, if injured or out of order, may represent a direct or indirect cause of accidents, release of toxic materials, or environmental contamination. These structures, systems, and components must be classified as safety-related and designed in such a way that remain undamaged and functional during and after major earthquakes.

C. Definition of the Design Earthquake

The evaluation of the design earthquake is the first step of the process that allows to design the structures, systems, and components of the plant in an earthquake-resistant way and to verify the seismic stability of the foundation soil. The design earthquake can be represented by seismic forces or by a design ground motion.

The seismic forces are directly evaluated making use of standard procedures, as those defined in the seismic building codes. In this case no dynamic response analysis is necessary and the evaluated pseudo-static forces are directly used, in combination with other loads, for design and verification.

The design ground motion is obtained using the seismic building code procedures or by investigations and analyses specifically carried out for the plant site. Dynamic response analyses are performed, in this case, in order to evaluate the seismic forces on the structures, systems, and components. The design ground motion is defined by a representative set of strong motion records or by response spectra and a duration.

The response spectra are particularly useful for the dynamic response analysis of the structures. They are curves, generally on a tripartite logarithmic graph paper, showing the variation of the peak responses (acceleration, velocity, and displacement) of a series of simple harmonic oscillators of different natural frequency and damping, subjected to the ground motion (Figure 5).

One of the most widely used definitions of duration is the time interval between the first and last peak of the ground motion above a specified amplitude level (5% of g is generally used for acceleration time-histories).

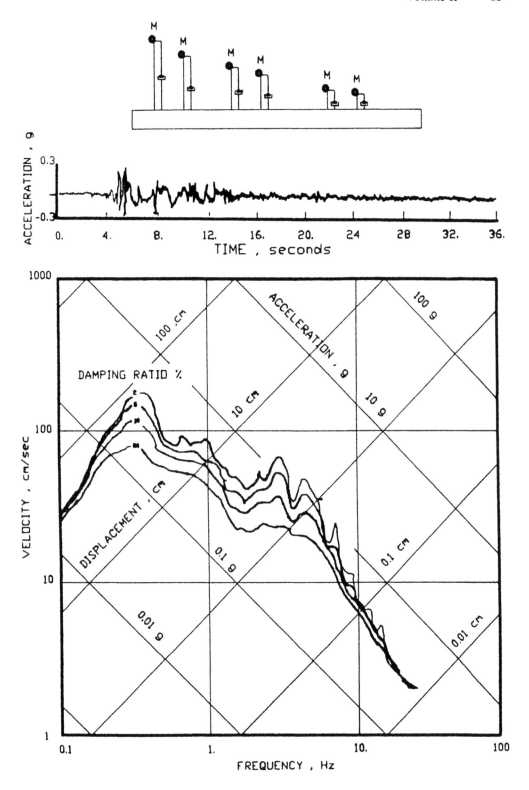

FIGURE 5.   Example of response spectra derived from 1980 Irpinia, Italy, earthquake, recorded at Sturno, E. W. component.

The design ground motion is defined at the surface and far from the buildings (free field) in order to distinguish how it is defined from how it is filtered through the soil-structure model and the structure-system/component models.

In engineering practice there are essentially three ways to evaluate the design earthquake:

1.   Standard procedures proposed by the seismic building codes
2.   Quasi-deterministic methodology
3.   Probabilistic methodology

1. Seismic Building Code Procedures

Seismic building code procedures are based on seismic hazard maps of the regions, which they are applicable to. These maps divide the region into different zones. Each zone is characterized by a seismic zoning parameter (zoning factor or zoning peak ground acceleration), representative of the seismicity of the zone.

The zoning factor, together with other coefficients, which account for the foundation soil properties and the building structural characteristics, allows to evaluate directly the seismic forces acting on the structures.[17] The zoning peak ground acceleration is used to scale standard response spectral shapes, proposed, in some cases, according to different local conditions.[18]

The seismic building codes are mainly devoted to design conventional buildings in such a way that, in order to prevent loss of lives, they resist severe earthquakes, without collapse, but with some structural and nonstructural damage. This criterium, generally, does not completely satisfy the safety requirements of major hazard industrial plants. Therefore, the building code requirements represent a minimum for design to be suitably integrated according to the potential hazard of the plant.

2. Quasi-Deterministic Method

The quasi-deterministic method evaluates the design ground motion on the basis of site-specific geologic, seismologic, and geotechnical investigations and analyses. It was developed looking mainly at the safety requirements of nuclear power plants, but it can be easily transferred to major hazard industrial facilities.

The method goes on through the following steps.

1.   The geologic and seismologic environment, in which the site is located, is defined.
2.   All the potential seismic sources (faults and seismogenic zones), important for the site, are characterized.
3.   The closest approaches between the seismic sources and the site are evaluated.
4.   The seismic characters of the potential earthquakes, associated to each seismic source, are estimated. Seismic character means all the available information regarding the historical seismicity: intensity, felt area, or the source mechanism: magnitude, hypocentral depth, and type of dislocation. Magnitude is the parameter most widely used to represent the seismic character.
5.   A recurrence relationship is developed for each seismic source. The recurrence relationship may be a key element for the design of major hazard industrial plants other than nuclear power plants. While the design practice generally adopted for nuclear facilities refers to the maximum credible earthquake associable to each seismic source, for the other industrial plants, the selection of the reference earthquake can be related to a recurrence relationship. For each seismic source the recurrence relationship is evaluated using data from historical seismicity, including instrumental data, or from geologic data, according to the following expression:[19]

$$\log n(M) = a\text{-}b\,M \tag{3}$$

Table 3
AMPLIFICATION FACTORS FOR
HORIZONTAL, RESPONSE SPECTRUM

Damping ratio (%)	Median			+ SD Median		
	a	v	d	a	v	d
0.5	3.68	2.59	2.01	5.10	3.84	3.04
1	3.21	2.31	1.82	4.38	3.38	2.73
2	2.74	2.03	1.63	3.66	2.92	2.42
3	2.46	1.86	1.52	3.24	2.64	2.24
5	2.12	1.65	1.39	2.71	2.30	2.01
7	1.89	1.51	1.29	2.36	2.08	1.85
10	1.64	1.37	1.20	1.99	1.84	1.69
20	1.17	1.08	1.01	1.26	1.37	1.38

[a] a = acceleration, v = velocity, and d = displacement.

From Newmark, N. M. and Hall, W. J., Development of Criteria for Seismic Review of Selected Nuclear Power Plants, Report NUREG/CR-0098, U.S. Nuclear Regulatory Commission, Washington, D.C., 1978.

where n(M) is the number of earthquakes per unit time, generally per year, with seismic character; for example, magnitude, greater than or equal to M and a and b are constants, determined on a case-by-case basis.

6. A reference earthquake, evaluated according to the previously developed recurrence relationship, is associated to each seismic source.

7. The geotechnical properties of the foundation soil at the site are investigated and their effects on the ground motion at the surface are analyzed.

8. The site ground motions at the surface, caused by each reference earthquake, associated to each seismic source, are estimated. For each seismic source, having known the closest approach to the site, the seismic character — magnitude — of the associated reference earthquake (maximum credible earthquake, or earthquake with a given recurrence) and, finally, the foundation soil properties, the induced site ground motion at the surface is evaluated. Different methods are available for the definition of the site ground motions. All are based on strong motion records, obtained during past earthquakes. Strong motion records, obtained at recording stations with soil condition similar to that of the plant site, during earthquakes with seismic character and distance from the site similar to those of the proposed reference one, can be selected in order to characterize the site ground motion. From the selected strong motion records, site response spectra — horizontal and vertical — are then statistically developed: mean and mean-plus-a-standard-deviation response spectra. As an alternative, peak ground acceleration, velocity, and displacement, for the appropriate site condition, magnitude, and distance from the source can be evaluated by attenuation relationships, such as those suggested in Tables 1 and 2. Site horizontal response spectra are then developed according to a standard procedure,[11] scaling the evaluated peak ground motion values by the amplification factors reported in Table 3 for different values of the damping ratio and for two levels of lognormally distributed probability. The suggested acceleration amplification factors are applied for frequencies up to 8 Hz,[20] while an acceleration amplification factor linearly varying to one on the tripartite logarithmic diagram is considered in the range 8 to 33 Hz. The response spectrum for the vertical component is, finally, determined scaling the horizontal spectrum by two thirds. An example of application of the described procedure is presented in Figure 6. Site response

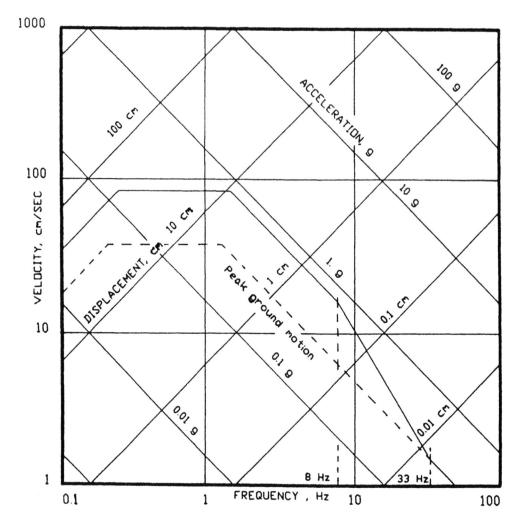

FIGURE 6.   Example of a site-specific horizontal response spectrum (0.3 g peak ground acceleration, soil condition, mean-plus-SD, 5% damping ratio), determined according to the Newmark and Hall procedure. (From Newmark, N. M. and Hall, W. J., Development of Criteria for Seismic Review of Selected Nuclear Power Plants, Report NUREG/CR-0098, U.S. Nuclear Regulatory Commission, Washington, D.C., 1978.)

spectra can be obtained, also, from attenuation relationships, which provide spectral ordinates as a function of magnitude, distance, and site condition.[21,22] Finally standard spectral shapes[23] (Figure 7) can be used. They are scaled at the peak ground acceleration value, determined by acceleration attenuation relationships, as those presented in Table 1. Like response spectra, duration is evaluated by representative strong motion records or by available correlations.[24,25]

9.   The design ground motion is finally defined on the basis of the site ground motions, accounting also for the design methods and the characteristics of the structures.[12] (The design ground motion is generally represented by smoothed large band response spectra.)

3. Probabilistic Method

The probabilistic methodology evaluates the level of the design ground motion that has a well-defined probability to be exceeded in a specified time interval. The design ground motion level is defined either by only one representative parameter, such as the peak ground acceleration (in this case standard spectral shapes are used to characterize the frequency

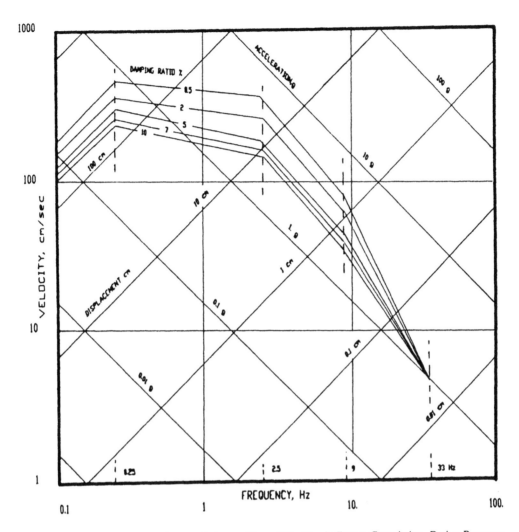

FIGURE 7. Standard horizontal spectral shapes. (From U.S. Atomic Energy Commission, Design Response Spectra for Seismic Design of Nuclear Power Plants, Regulatory Guide 1.60, U.S. Atomic Energy Commission, Washington, D.C., 1973.)

content), or by uniform hazard response spectra (horizontal and vertical), i.e., response spectra which have a uniform probability of exceedence in the time interval of interest.

As in the quasi-deterministic method, the geologic and seismologic environment of the site shall be defined and the seismogenic sources shall be characterized. After that, the annual probability λ_n, that a given ground motion parameter (peak ground acceleration or response spectra ordinate), due to a particular seismic source, will be exceeded at the site, is evaluated convoluting the following probability functions:[26]

1.   The probability that an earthquake of given magnitude will occur on a given seismic source during a specified time interval. (It is evaluated by a recurrence relationship, previously developed for the seismic source of interest.)
2.   The probability that the energy release on the seismic source is at a given distance from the site. (It is assessed considering both the geometry of the seismic source, e.g., the fault length, and the relationship between magnitude and fault rupture length, or rupture area.[27,28])

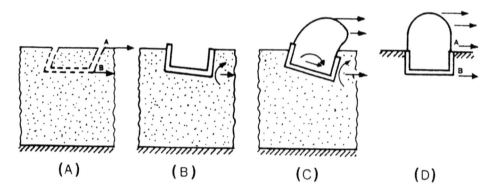

FIGURE 8.   Soil-structure interaction effects: schematic situations.

3.   The probability that the site ground motion from an earthquake of a given magnitude, occurring at a given distance, exceeds a defined level. (It is estimated by selected attenuation relationships: magnitude-distance-peak ground motion values[7] or magnitude-distance-spectral ordinates.[21,22])

If there are n seismic sources, the total annual proability λ is computed simply adding the contribution λ_n from each source. The inverse of λ is the return period in years.

Finally, the probability that a selected design ground motion level is exceeded in a selected time interval t (in years) is given by:

$$P = 1 - \exp(-\lambda t) \qquad (4)$$

D. Modeling of Soil, Structure, Systems, and Components
1. Soil-Structure Interaction

The dynamic interaction of the structures with the foundation soil (soil-structure interaction), in some cases, significantly affects the structural response.

Soil-structure interaction changes the characteristics of the free-field motion at the foundation depth; in other words, there are soil-structure interaction effects when the motion at the interface between soil and structure is different from the motion that would occur if the structure were not present.

To explain the above-mentioned effects the schematic situations shown in Figure 8 are compared.

In Figure 8A the dynamic response of a soil layer overlying a bedrock, due to vertically propagating shear waves, is considered. As previously discussed, in a free-field situation the seismic motion, propagating from the bedrock up to the surface, changes characteristics. The ground motion at the surface is affected by the natural frequencies of the soil layer and lead to higher horizontal displacements than those produced in depth.

If the soil is excavated, as shown in Figure 8B, and a rigid boundary without mass is inserted inside, the motion at points A and B is different from that which would occur in the free-field situation of Figure 8A. In fact, the presence of the excavation will cause wave reflection and scattering and the rigid boundary will average the displacements between the surface and the base and will experience a rigid body displacement with translational and rocking components. This effect is known as "kinematic interaction". If the hypothesis of vertically propagating shear waves is assumed, the kinematic interaction exists only in the case of embedded foundations. For surface foundations, the kinematic interaction may be important only if surface waves or inclined shear waves are considered.

Finally, the structure with its mass, placed inside the excavation (Figure 8C), will generate inertial forces that will lead to an overturning moment and a base shear. These will cause deformations in the foundation soil, modifying, again, the motion at the base of the structure. The aforesaid effect is called "inertial interaction"; it coincides with the effect due to dynamic loads directly applied to the structure.

If the same problem is considered for a structure founded on rock (Figure 8D), no appreciable amplification effects will be recorded between points B and A and the acceleration will be practically the same along the embedded sides of the structure. Moreover the overturning moment and the base shear, due to the structure vibrations, will not induce significant strains in the rock. The motion at the base of the structure, in this case, is practically the same of the free-field surface motion.

Summing up, it is possible to conclude that in many cases (light structures, surface foundations, stiff soil, or rock) modifications to the free-field motion, due to soil-structure interaction, are practically equal, if not less, to the uncertainties related to the definition of the design ground motion; the soil-structure interaction effects may contribute only for a few percent to the deformation of the structure. In these cases the interaction effects are neglected and fixed base structural models are considered for design. This is the choice generally adopted by the seismic building codes. In other cases (rigid and massive structures, embedded foundations, soft soil) modifications to the free-field motion are important both for the evaluation of the dynamic forces applied to the structure and for the definition of the vibratory motions at the different building floors used in their turn, for the design of the plant systems and components.

Fixed base analyses generally lead to larger values of the global response (overturning moment and base shear) than those obtained from a soil-structure interaction analysis, while the displacements at the top of the structure may be larger when the soil flexibility is accounted for.

Moreover, about the evaluation of the motion at different floors, the following differences with a fixed base analysis should be considered:

- Sliding and rocking, due to kinematic interaction, modify the frequency content of the motion at the floors
- The presence of a flexible constraint (the soil) at the base of the structural model, decreases the natural frequencies of the structure, affecting the transfer function of the motion
- The vibration of the structure generates waves, which, propagating through the soil, cause a decrease of the energy density with the distance from the source of vibration. The decrease of energy density leads to a damping effect (the geometric damping) on the dynamic behavior of the structure. For a soil approaching an elastic half-space the geometric damping is high, while, for a soil layer overlying a bedrock, it is practically zero for frequencies of vibration less than the natural frequency of the layer. In this case, in fact, the propagating energy is entrapped inside the soil layer.

2. Modeling of Soil

Two different methods are available to model the soil region which contributes to soil-structure interaction: the direct method[29] and the three-steps method.[30] Both take into account the nonlinear soil behavior (the shear modulus and the material damping of the soil are functions of the shear strain) by an equivalent-linear analysis technique.[31] This iterative procedure adequates the elastic modulus and the material damping of the soil to the shear strain evaluated in the previous iteration. Therefore, in each iterative step, a linear analysis with effect superposition is carried out.

In the direct method the soil and structure are modeled together using the finite element technique. In modeling the soil, two different needs have to be matched: (1) the modeled

Table 4
**SPRING CONSTANTS AND GEOMETRIC DAMPING RATIOS FOR
SURFACE FOUNDATIONS ON AN ELASTIC HALF-SPACE**

Mode of vibration	Equivalent radius (R)	Spring constant (k)	Inertia ratio (b)	Damping ratio (β_G)
Vertical	$(BL/\pi)^{0.5}$	$\dfrac{4GR}{1-\nu}$	$\dfrac{m(1-\nu)}{4\rho R^3}$	$\dfrac{0.425}{b^{0.5}}$
Sliding	$(BL/\pi)^{0.5}$	$\dfrac{8GR}{2-\nu}$	$\dfrac{m(2-\nu)}{8\rho R^3}$	$\dfrac{0.288}{b^{0.5}}$
Rocking	$(BL^3/3\pi)^{0.25}$	$\dfrac{8GR^3}{3(1-\nu)}$	$\dfrac{3I_x(1-\nu)}{8\rho R^5}$	$\dfrac{0.15}{(1+b)b^{0.5}}$
Torsional	$[BL(B^2+L^2)/6\pi]^{0.25}$	$\dfrac{16GR^3}{3}$	$\dfrac{I_z}{\rho R^5}$	$\dfrac{0.5}{1+2B}$

soil region must have limited dimensions and (2) the propagation of waves through a semi-infinite medium has to be accounted for.

For this reason special boundary conditions are applied to the soil model to take into account the outward wave propagation (consistent boundary) and to take up energy (viscous boundary); the last is generally used to simulate three-dimensional effects.

In the direct method the dynamic analysis of the whole soil-structure model is preceded by a deconvolution analysis, in order to transfer the design ground motion from the free-field surface — where it is defined — to the base of the model. Then, the soil-structure interaction analysis allows to evaluate, in the same time, the dynamic response both of the soil and of the structure.

If the structure is very complex, it cannot be modeled in an accurate way by the direct method. In this case, the interaction analysis is limited to evaluate the motion at the base of a simplified structural model. Then the evaluated base motion is applied to a more accurate fixed base model of the structure.

In the three-step method the analysis is shared into three phases.

Kinematic interaction — Starting from the design ground motion at the free field surface, the motion at the base of the structure is evaluated. Assuming that the foundation is rigid, only a rigid massless interface is necessary to characterize the presence of the structure. The base motion is, in this case, a rigid body motion with six components (three translational and three rotational). If the foundation is flexible, it is necessary to evaluate the base motion in different points along the interface between the soil and the structure. For surface foundations, assuming vertically propagating shear waves, the kinematic interaction effect does not exist and this step is not considered.

Evaluation of soil impedance functions — The soil impedance functions are complex functions of frequency and represent the soil response in terms of forces and moments, applied to the structure, when a unit displacement or rotation, sinusoidally varying in time, is applied to the soil. The real and imaginary parts of the function represent, respectively, the stiffness and the geometric damping of the soil. Many solutions have been proposed to account for different geometric situations of the foundation and different soil behaviors.[32,33] Simplified solutions, that allow to represent the soil with springs and dashpots with frequency-independent properties, are also available.[32-34] Table 4 reports expressions for the frequency-independent spring constants k and the geometric damping ratios β_G, according to different vibration modes, for a circular surface foundation on an elastic half-space. R is the radius of the foundation, m is the mass, I_x and I_z are, respectively, the mass moment of inertia of the foundation about the horizontal and vertical axes. G, ρ and ν are, respectively, the elastic shear modulus, the mass density, and the Poisson ratio of the soil. For real soil the material

Table 5

SPRING CONSTANTS FOR FOUNDATIONS EMBEDDED IN AN ELASTIC LAYER OVERLYING A HALF-SPACE

Mode of vibration	Spring constant
Vertical	$\dfrac{4GR}{1-\nu}\left(1+1.28\dfrac{R}{H}\right)\left(1+0.47\dfrac{E}{R}\right)\left[1+\left(0.85-0.28\dfrac{E}{R}\right)\dfrac{E/H}{1-E/H}\right]$
Sliding	$\dfrac{8GR}{2-\nu}\left(1+\dfrac{1}{2}\dfrac{R}{H}\right)\left(1+\dfrac{2}{3}\dfrac{E}{R}\right)\left(1+\dfrac{5}{4}\dfrac{E}{H}\right)$
Rocking	$\dfrac{8GR^3}{3(1-\nu)}\left(1+\dfrac{1}{6}\dfrac{R}{H}\right)\left(1+2\dfrac{E}{R}\right)\left(1+0.7\dfrac{E}{H}\right)$
Torsional	$\dfrac{16GR^3}{3}\left(1+2.67\dfrac{E}{R}\right)$

damping ratio should be added to the geometric damping ratio. For rectangular surface foundations the same expressions can be utilized, determining an equivalent radius, as also shown in Table 4 (B is the width of the foundation, along the axis of rotation for rocking modes, and L is the length, in the plane of rotation for rocking modes). For foundations embedded in a homogeneous elastic soil layer overlying a rigid half-space, the expression for the spring constants[35,36] are reported in Table 5 (E is the depth of embedment and H is the thickness of the layer).

Inertial interaction — In the last step of analysis, the motion, defined at the soil-structure interface, according to point a), is applied to a dynamic model of the soil-structure system, where the soil is modeled by impedance functions (or, in the simplified procedure, by springs and dashpots), evaluated according to point b).

It should be noted that, in engineering practice, the soil-structure interaction analysis is generally limited to the inertial interaction of simplified models.

The soil properties, modeled by springs and dashpots, are varied parametrically and the sensitivity of the dynamic response is evaluated for extreme cases.

For surface foundations on soil deposits approaching to a half-space, the approximation is acceptable. Conservativism should be applied, however, in defining the soil spring constants and the geometric damping ratios, especially when the half-space behavior is a rough hypothesis.

For embedded foundations, neglecting the kinematic interaction generally may lead to conservative results.[37] However, when it is necessary to evaluate not only the global dynamic behavior but the motion at the floors as well, the conservativism of the design method of analysis has to be checked by sensitivity analyses, using the direct method or the complete three-step method and simplified structural models.

3. Modeling of Structures

In defining the structural model it is necessary to match the need of developing dynamic models with a limited number of degrees of freedom with the necessity of suitably describing the structure and the material behavior. Particular care should be taken of the mass and the stiffness distribution and of the dissipative (damping) and nonlinear (ductility) behavior of the structure.

The structural model may be defined both by finite elements and by a lumped-mass technique.[38] These two approaches are quite similar: both discretize the physical properties of the structure, lumping the masses at discrete nodes, connected each other with massless elements, having appropriate stiffness and damping properties.

The main difference is that the lumped-mass technique discretizes the structure on a larger scale; for a structure with most of the masses concentrated at the floor levels, the node coordinates are chosen such as to represent these levels.

Table 6
SUGGESTED STRUCTURAL DAMPING RATIOS

Stress level	Type and condition of structure	Damping ratio (%)
Working stress no more than about 1/2 yield point	Vital piping	1—2
	Welded steel, prestressed concrete, well-reinforced concrete (only slight cracking)	2—3
	Reinforced concrete with considerable cracking	3—5
	Bolted and/or riveted steel, wood structure with nailed or bolted joints	5—7
At or just below yield point	Vital piping	2—3
	Welded steel, prestressed concrete (without complete loss in prestress)	5—7
	Prestressed concrete with no prestress left	7—10
	Reinforced concrete	7—10
	Bolted and/or riveted steel, wood structures with bolted joints	10—15
	Wood structures with nailed joints	15—20

From Newmark, N. M. and Hall, W. J., Development of Criteria for Seismic Review of Selected Nuclear Power Plants, Report NUREG/CR-0098, U.S. Nuclear Regulatory Commission, Washington, D.C., 1978.

The finite element technique allows a more detailed description. Different finite elements can be used and different formulations are adopted in order to represent the discretized masses (lumped and consistent mass).

In discretizing the structure, lumping of masses has to be performed without any total mass variation, maintaining the center of gravity and the rotational inertia for the whole structure for each important system and component, housed inside.

Even if it is possible to analyze three-dimensional models, when the structural responses are uncoupled for orthogonal directions, it is preferable to consider three separated two-dimensional models. The dynamic responses, obtained according to three orthogonal components of the design ground motion, are then superimposed.

Regarding the dissipative behavior of the structure, it should be emphasized that the choice of appropriate values for the structural damping[11,39] is a very critical step. In fact damping influences the results of the dynamic analysis in a remarkable way. The structural damping ratio, defined as percentage of the critical damping, takes into account the dissipation of energy due both to the material type and to the structural element type. Moreover, it varies according to the stress level induced by the earthquake. Suggested values of the structural damping ratio are presented in Table 6.

Soil-structure interaction analyses, developed according to the modal analysis technique (as it will be seen in the following), need to refer, for each vibration mode, to a unique damping value of the overall soil-structure system: the weighted modal damping.[40,41]

The dynamic response of the structures involves a nonlinear behavior, according to an elastoplastic resistance-displacement relationship. In some cases, it may be convenient to account for the capacity of the structures to deform beyond the elastic limits, remaining, however, within acceptable plastic limits and without loss of strength. The ductility factor represents this capacity of deformation and it is defined by the ratio between the allowable plastic displacement and the elastic one.

Each structural element has its own ductility factor, generally different from the ductility factors of the others, and contributes to the overall ductility of the structure.

The structural nonlinear behavior can be accounted for, in a simple way, defining an overall ductility factor for the structure and carrying out a linear dynamic analysis, where the input motion is represented by inelastic response spectra, obtained by modifying the design response spectra by the overall ductility factor.[42]

For very critical structures, such as nuclear power plant buildings, the nonlinear behavior is conservatively neglected for design purposes and it is requested that the dynamic response remains within elastic limits.

4. Modeling of Systems and Components

The plant systems and components, which can be uncoupled[43] from the supporting structure (also called "primary system"), are selected and defined as "secondary systems".

The systems, having stiffness and mass such to modify the dynamic response of the primary system, have to be included with their mass and stiffness in the primary model.

The system with relatively small mass and high natural frequency generally have a negligible interaction with the structure. They can be uncoupled from the structure, including only the mass in the primary model. The secondary systems are then analyzed, after the structure, considering as input motion the vibratory motion at the floor of the building, on which they are housed. The motions at the floors are obtained from dynamic analyses of the primary model. They are represented by "floor response spectra", which define the peak response of a series of simple harmonic oscillators of different natural frequency and damping, located at the support of the secondary system, when the primary one is shaked by the design ground motion.

To take into account the uncertainties in the material properties of the structure and the soil as well as the approximations in modeling, the computed floor response spectra are smoothed and the peaks are broadened according to suggested criteria.[44]

5. Modeling of the Dynamic Effects of Fluids

Hydrodynamic effects need to be considered when structures containing fluids (tanks) are present in the plant layout.

During earthquakes the fluid develops a hydrodynamic pressure on the container. This pressure can be separated into two different components: the impulsive and the convective.

The impulsive pressure is generated by a portion of fluid which acts as rigidly connected to the tank. The corresponding force is equal to the mass of this portion of fluid multiplied by the acceleration of the tank.

The convective component is determined by the fluid oscillations. The corresponding force is evaluated considering a portion of fluid as an oscillating mass flexibly connected to the tank.

The dynamic model of the tank with the fluid can be represented by two masses: (1) the mass of the tank plus the mass of the rigidly connected portion of fluid and (2) the oscillating fluid mass, which is connected to the tank walls by elastic springs.

The inertia, the geometry (location of the masses above the base of the tank), and the elastic properties of the dynamic model are defined as a function of the geometry of the tank and the level of the fluid inside.[45]

If the tank is supported by a flexible structure (elevated tanks), the dynamic behavior of the supporting structure shall be included into the model.

E. Evaluation of the Seismic Response

The seismic response of the structures, systems, and components of the plant can be obtained with two different techniques of analysis: dynamic methods[38] and pseudo-static methods.[17] Both procedures allow to evaluate the seismic forces on the structures and the systems.

The dynamic methods utilize the design ground motion (design response spectra or representative acceleration time histories) and the model of the structure (a fixed base model or a soil-structure interaction model) as input data for the analysis. The results of the dynamic analyses are generally presented in terms of diagrams of the inertial forces along the structure and in terms of floor response spectra.

The inertial forces, combined with the other loads, allow to design and verify the structures.

The floor response spectra are utilized for the evaluation of the seismic response of the secondary systems. For this purpose both analytical methods (dynamic methods and pseudo-static methods) and qualification tests are employed.

The pseudo-static methods apply both to the structures and to the secondary systems. They evaluate the inertial forces as equivalent static forces. For the structures, the base shear is determined as a percentage of the total weight. The used percentage is a function of many parameters, which are determined on the basis of the seismic building codes, according to the regional seismicity, the characteristics of the foundation soil, and the type and natural frequency of the structure. The inertial forces are then determined from the base shear according to a given distribution rule along the height of the structure.

For the secondary systems the inertial force, applied at the center of gravity, is evaluated as a percentage of the weight of the system. The adopted percentage is determined from the pertinent floor response spectra or on the basis of building code procedures.

1. Dynamic Methods

The equation of motion for a dynamic system with n degrees of freedom, subjected to an acceleration time history at the base, is given by:

$$[M] \{\ddot{x}\} + [C] \{\dot{x}\} + [K] \{x\} = -[M] \{\ddot{u}\} \tag{5}$$

where [M] is the mass matrix, [C] is the damping matrix, and [K] is the stiffness matrix of the dynamic system. The vectors $\{x\}$, $\{\dot{x}\}$, and $\{\ddot{x}\}$ are the relative-to-the-base displacements, velocities, and accelerations of the n degrees of freedom. The vector $\{\ddot{u}\}$ represents the base acceleration (note that the absolute accelerations of the n degrees of freedom are $\{\ddot{z}\} = \{\ddot{x}\} + \{\ddot{u}\}$).

The equation of motion (Equation 5) can be solved by direct integration or by the modal analysis technique. This last is widely used for dynamic analysis and will be described in the following.

The modal analysis allows to evaluate the dynamic response of the structures in their linear range of behavior.

It is essentially based on the fact that, for the damping models used for structural systems, the dynamic response for each natural mode of vibration can be computed independently of the others, and the total response can be obtained combining the modal responses. In other words, the dynamic analysis of a multi-degree-of-freedom-system is substituted by the dynamic analysis of n single-degree-of-freedom systems, each relative to a mode of vibration. Generally it is unnecessary to determine the contribution of all the n modes of vibration, because, in seismic analyses, the response is primarily due to the first modes of vibration.

If the undamped free vibration condition for the dynamic system is considered, Equation 5 reduces to:

$$[M] \{\ddot{x}\} + [K] \{x\} = 0 \tag{6}$$

operating the following linear transformation for the coordinates:

$$\{x\} = [\Phi] \{q\} \tag{7}$$

where $\{q\}$ is of the form $\{e^{i\omega t}\}$, Equation 6 becomes:

$$([K] - [\omega_j^2] [M]) [\Phi] \{q\} = 0 \tag{8}$$

Solutions for Equation 8 are given by the n eigenvectors $\{\Phi_j\}$, corresponding to the n eigenvalues ω_j^2. ω_j is the natural circular frequency of the jth mode of vibration; $\{\Phi_j\}$ is the modal shape, i.e., the shape of deformation related to the jth mode; and q_j is the generalized coordinate which represents the time-varying amplitude of vibration of the jth mode.

Applying the same linear transformation of coordinates (in Equation 7) to the equations of motion (in Equation 5) and assuming that [C] can be expressed by a combination of [M] and [K], the matrices [M], [K], and [C] are transformed in diagonal matrices and the resulting equation becomes uncoupled, i.e., each equation refers to only one generalized coordinate.

With the following position:

$$\{\ddot{u}\} = \{\delta\} \ddot{u}(t) \tag{9}$$

where $\{\delta\}$ is a unit vector with a one for the degrees of freedom corresponding to the direction of the base motion $\ddot{u}(t)$, and a zero for all the other degrees of freedom, the jth equation of motion becomes:

$$\ddot{q}_j + 2\beta_j \omega_j \dot{q}_j + \omega_j^2 q_j = -\gamma_j \ddot{u}(t) \tag{10}$$

where β_j is the damping ratio for the jth mode and γ_j is the modal participation factor:

$$\gamma_j = \frac{\{\Phi_j\}^T [M] \{\delta\}}{\{\Phi_j\}^T [M] \{\Phi_j\}} \tag{11}$$

Having known the acceleration time history $\ddot{u}(t)$, Equation 10 can be solved by applying a convolution integral (Duhamel integral) and determining, for each vibration mode, the generalized coordinate $q_j(t)$.

The displacement relative to the base, for each degree of freedom i of the dynamic system, is, then, obtained from the first m significant mode of vibration, by:

$$x_i(t) = \sum_j^m \Phi_{i,j} q_j(t) \tag{12}$$

Moreover, it can be demonstrated that the absolute acceleration of the ith degree of freedom is given by:

$$\ddot{z}_i(t) = -\sum_j^m \omega_j^2 \Phi_{i,j} q_j(t) \tag{13}$$

A complete description of the response history is necessary when the dynamic analysis is performed in order to evaluate the floor response spectra at the different levels of the structure. When the analysis is limited to compute the inertial forces acting on the structure or a secondary system, then only the maximum responses are determined. In this case, the response spectra represent useful tools for characterizing the design ground motion or the motions at the different floors of the structure. In fact, when a series of single-degree-of-freedom systems are subjected to a base motion $\ddot{u}(t)$, the corresponding response spectrum gives, for each system, the maximum response (e.g., the spectral displacement Sd_j) knowing the natural frequency ($f_j = \omega_j/2\pi$) and the damping β_j of the system.

Equation 10 is the equation of motion of a single-degree-of-freedom system, subjected to the base motion $\gamma_j\ddot{u}(t)$. Therefore the maximum displacement for this system, i.e., the generalized coordinate for the jth mode, is given by:

$$q_j = \gamma_j Sd_j \tag{14}$$

The maximum relative-to-the-base displacements of the dynamic system, due to the jth mode of vibration, are then:

$$\{x_j\} = \{\Phi_j\}\,\gamma_j Sd_j \tag{15}$$

The displacement vector $\{x_j\}$ represents the modal response. It leads directly to other modal quantities, as the modal maximum inertial forces:

$$\{F_j\} = [M]\,\{\Phi_j\}\,\gamma_j Sa_j \tag{16}$$

where $Sa_j = \omega_j^2 Sd_j$ is the spectral acceleration.

The computed modal responses $\{R_j\}$ (displacements, inertial forces, shears forces, moments, stresses) are, finally, combined[46] taking into account that the maximum responses in all modes do not occur normally at the same time, but with a phase difference.

For this reason, when the modes of vibration are not closely spaced in the frequency range (two modes are considered closely spaced if their frequencies differ from each other by 10% or less of the lower frequency), the square-root-sum-of-squares rule is applied to evaluate the response of the ith degree of freedom:

$$R_i = \left(\sum_1^m{}_j R_{i,j}^2\right)^{0.5} \tag{17}$$

where m is the number of significant modes.

If a few modes are closely spaced, the combined response is obtained by computing the absolute sum of the closely spaced modal responses and then combining the result with the other modal responses by the square-root-sum-of-squares rule.

Normally the response R_i is obtained considering only one of the three components of the design ground motion. Therefore, the responses, due to each base motion component, shall be computed and finally combined. The square-root-sum-of-squares rule is applied also in this case,[46] to take into account that the contributions to the total response of the ith degree of freedom will not occur at the same time.

2. Pseudo-Static Methods

The pseudo-static methods can be used for buildings having regular distribution of stiffness and mass over height. The seismic response is directly evaluated in terms of equivalent static forces distributed over the height of the structure. The capacity of the structure to deform beyond the elastic limit without loss of strength (ductility) is normally accounted for.

The pseudo-static methods are recommended for conventional buildings, which are requested to resist major earthquakes, without collapse, but with limited structural and non-structural damage. Extention of the methods to major hazard industrial plants should be considered on a case-by-case basis, modifying, if necessary, the values of the coefficients which contribute to the definition of the equivalent static forces.

Different versions of the method are available in different countries, however it is possible to recognize that, in general, they are based on the definition of the base shear according to the following:[17]

$$V = ZIKCSW \tag{18}$$

where Z is the seismic zoning factor. It is related to the seismicity of the region and it is determined on the basis of seismic hazard maps. I is the occupancy importance coefficient. It has been introduced to provide a higher level of seismic protection for those buildings or facilities which are requested to remain functional for past earthquake operations (such as hospitals, fire fighting stations, or electric power supply facilities) or buildings which are likely to be particularly crowded during earthquakes (such as schools or public buildings). K is the structural coefficient. It depends on type and characteristics of the structural system. For types of construction, which are recognized as having a good performance during earthquakes (higher ductility), the lower K values are considered; on the contrary, the higher K values are assigned to types of construction which are highly vulnerable to seismic forces. Moreover, it should be noted that the K values applied to conventional buildings are lower than those assigned to other construction types, because conventional buildings have non-structural and noncomputed resisting elements which increase resisting capacity and ductility. C is the resonance coefficient. It represents the combined response due to all the vibration modes and is expressed as a function of the fundamental natural period of vibration T of the structure (that is the inverse of the first natural frequency). The period T is computed considering the properties and the deformational characteristics of the structure, or using suggested relationships to the dimensions of the building. Different formulas have been proposed by the different national building codes in order to evaluate the fundamental natural period of vibration and to compute the resonance coefficient. S is the soil-structure resonance coefficient. It accounts for the local seismic response of the foundation soil in relation to the structural response, that is, for the fact that greater structural damage is likely to occur when the fundamental natural period of the structure is similar to that of the foundation soil deposit. The coefficient S is evaluated as a function of the ratio between the fundamental natural period of the structure T and that of the soil deposit T_s. The lower S values are considered when the ratio T/T_s is far from unity, while the highest S value is assumed when the ratio T/T_s is equal to one. Some building codes suggest the S value only on the basis of a qualitative description of the foundation soil. In some cases, the coefficient S accounts also for the type of the foundation structure. W is the total weight of the structure, plus all the dead loads (nonstructural elements, piping, equipment), plus that percentage of the live loads, which reasonably may be present during the earthquake.

The lateral forces at different levels (floors) are finally determined using the following ''triangulare distribution'' rule:

$$F_i = \frac{V w_i h_i}{\sum_1^n w_i h_i} \tag{19}$$

where w_i is the weight assigned at the ith level and h_i is the height of the ith level above the base.

Equation 19 is similar to the expression of the modal lateral forces that could be determined with a modal dynamic analysis performed on a shear-type model with lumped masses w_i/g at each floor:[38]

$$F_{i,j} = \frac{V_j w_i \Phi_{i,j}}{\sum_1^n w_i \Phi_{i,j}} \tag{20}$$

where V_j is the modal base shear and $\phi_{i,j}$ is the ith ordinate of the jth modal shape.

The two expressions coincide when the dynamic response is represented by the first mode of vibration and when the first modal shape is a straight line. For short period structures (low and rigid buildings) the triangular distribution rule represents a good approximation. For long period structures (high and flexible buildings), the first modal shape departs from the straight line with an increased deflection at the top and the contribution of the higher modes become significant. In this case Equation 19 is substituted by a similar one where, instead of the base shear V, the difference V-Ft is considered. Ft is a percentage of V, concentrated at the top of the structure.

Finally the effects of the lateral forces, assumed to act in two orthogonal directions, are determined from two independent analyses and are combined.

The pseudo-static methods can be used also for systems and components. If floor response spectra are available, the response of a secondary system can be evaluated, rather than with a dynamic analysis, considering the peak value of the spectral accelerations, related to a selected damping value, multiplied by an appropriate factor. This last accounts for the effects of multimode response (it is assumed equal to one, if the system can be adequately modeled as a single-degree-of-freedom system). The seismic force is, then, obtained by multiplying the mass of the system, concentrated at its center of gravity, by the factored peak spectral acceleration.

The seismic building codes allow, in some cases, to determine the seismic equivalent static forces on nonstructural elements and equipment, according to an expression similar to that suggested for structures:[17]

$$Fc = ZICpWp \qquad (21)$$

where Z is the seismic zoning factor, I is the importance coefficient for the nonstructural element, Cp is a function of the element type, and Wp is the weight.

F. Stability of Foundation Soil

Soil conditions have an important role on building damage during earthquakes. As described in a previous section, soil stratigraphy and properties influence the characteristics of the ground motion at the free-field surface. Moreover, the motion at the base of the foundations can be affected by soil-structure interaction phenomena.

Last, but not least, soil instability represents a considerable source of damage for buildings; in fact, large soil deformations and failure during major earthquakes are induced by the seismic waves which propagate through the soil, or by the combined effect of the static and dynamic loads transmitted to the foundation soil by the structures. Landslides could affect soil slopes, causing damage to buildings founded on them or located nearby. Retaining structures, finally, could slide or overturn because of the dynamic stresses applied by the soil.

1. Liquefaction

Liquefaction phenomenon is one of the main sources of soil instability during earthquakes. It has been observed in saturated cohesionless soil deposits, mainly loose sands, and, in some cases, resulted in severe damage to buildings (as during the Niigata, Japan, earthquake[47] of June 16, 1964) or major landslides (as that developed at Anchorage,[48] Alaska, during the earthquake of March 27, 1964).

Starting from 1964 scientific interest concentrated on the analysis of liquefaction and on developing engineering procedures in order to evaluate the potentiality that liquefaction could happen in an area of interest.

Detailed field observations and geotechnical investigations have been carried out at several sites, pointed to having experienced liquefaction during past earthquakes.[49-53] They dem-

Table 7
FACTORS AFFECTING LIQUEFACTION PHENOMENON

Factor classification	Liquefaction resistance	Earthquake-induced shear stresses
Soil properties	Grain size distribution, soil stucture, relative density	Elastic shear modulus, mass density
Environmental factors	Seismic history, geologic history (aging), lateral earth pressure coefficient, effective confining pressure	Topographic and stratigraphic characteristics
Earthquake characteristics	Equivalent number of representative cycles (earthquake magnitude)	Peak ground motion values, frequency content

onstrated that liquefaction is more likely to occur in deposits of recent origin (Olocene or late Pleistocene)[54] with the groundwater table near the surface, mainly constituted of medium and medium-fine loose sands with a uniform grain size distribution and low percent of fine.[55] The documented case histories of liquefaction and no liquefaction during past earthquakes represent a useful set of data in order to develop correlations between soil properties and liquefaction potential.[56]

In the laboratory, dynamic tests (cyclic triaxial and simple shear tests) on undisturbed or reconstituted soil samples have been prepared to reproduce site conditions during earthquakes.[57,58] Laboratory tests allowed a better understanding of liquefaction phenomenon, pointing out the specific factors that determine it.[59,60] These factors can be identified as soil properties, environmental factors and earthquake characteristics. Table 7 shows the main factors, whose combination determines the phenomenon, distinguishing those that influence liquefaction resistance from those that concern shear stresses induced by earthquakes.

Finally, theoretic studies have proposed soil constitutive laws that allow to explain field observations and laboratory test results.[61]

Liquefaction phenomenon can be explained as described in the following. Shear stresses, due to propagating shear waves, are induced in soil deposits during earthquakes. Because of these stresses, cohesionless soils tend to become more compact, reducing their volume. If the soil is saturated by water and its permeability, related to the time of application of the dynamic loads, is such that an undrained condition can be assumed (i.e., no pore water drainage is allowed), the tendency to volume reduction results in a pore water pressure build-up. If the pore water pressure approaches the applied confining pressure, the intergranular pressure (effective pressure) becomes zero and the soil loses its shear strength and behaves as a fluid.

If the soil is a loose sand, the pore water pressure quickly approaches the confining pressure and the soil experiences a permanent loss of strength with large deformations.

If the soil is a dense sand, even if a zero-effective pressure condition is reached, soil deformations stop at a threshold value. In this case, in fact, a dilatative behavior, typical of dense sands, causes a reduction of the pore water pressure and, at last, the soil can develop enough strength to withstand the applied shear stresses.

After the earthquake shaking, the pore water pressure build-up tends to dissipate, leading normally to an upward flow of water. This could cause liquefaction of the overlying layers because of a quick condition. Even if this last condition is not triggered, the consolidation of the sand layers will lead to volume reduction and settlements at the surface of the soil deposit.

Evaluation of liquefaction potential at a given site requires both a dynamic analysis of the soil deposit for a defined design earthquake and the evaluation of the soil liquefaction resistance.

The dynamic shear stresses, induced within the soil deposit by the design earthquake, are evaluated, in many cases, with a dynamic deconvolution analysis of the surface ground motion through a one-dimensional soil model.[16] The design earthquake, in this case, is represented by an ensemble of representative strong motion records.

As an alternative, a simplified procedure[62] can be used. In this case the average shear stress τ_{av} induced by the earthquake at a given depth h from the surface, is evaluated according to the following:

$$\tau_{av} = 0.65 \; \gamma h \; \frac{a}{g} \; r_d \qquad (22)$$

where a/g is the peak ground acceleration at the surface, expressed in g, γ is the unit weight of the soil and r_d is a stress reduction coefficient with a value less than one, which accounts for the deformability of the soil deposit (Figure 9).

The liquefaction resistance is evaluated both from cyclic laboratory tests and from *in situ* tests. Laboratory tests should reproduce the *in situ* soil conditions and state of stresses in order to provide reliable results. The difficulty both to obtain undisturbed sand samples and to realize a cyclic simple shear state of stresses is well known. Therefore, the results from standard cyclic tests (such as the cyclic triaxial tests), performed on reconstituted samples, are corrected by appropriate coefficients, which allow to lead back to the site conditions.[59]

In situ tests — standard penetration test (SPT) and cone penetration test (CPT) — allow to overcome most of the problems related to laboratory tests. They have received increasing attention recently and procedures, which provide good correlations between test results and liquefaction resistance, have been proposed.[56,63]

In particular the SPT is the *in situ* test which gives a soil index well correlated to those factors which influence the soil liquefaction resistance, such as relative density, soil structure, geologic and seismic history, and lateral earth pressure coefficient.[12] Moreover, as SPT is a widely used test, a great deal of SPT data, related to sand deposits which experienced or did not experience liquefaction during past earthquakes, are available. From these data the chart of Figure 10 has been developed in order to evaluate liquefaction resistance.[56] The cyclic stress ratio causing liquefaction (i.e., the ratio between the average shear stress τ_{av} and the effective overburden pressure σ'_o is determined as a function of the soil resistance index $N_{1(60)}$:

$$N_{1(60)} = C_N \; N \; ER/60 \qquad (23)$$

where N is the number of blows for foot, measured in the field during the test and C_N is a coefficient, function of the effective overburden pressure at the depth where the penetration test was conducted, which can be expressed by the following:[64]

$$C_N = (1/\sigma'_o)^{0.5} \qquad (24)$$

with σ'_o expressed in kg/cm² and ER/60 is a coefficient which normalizes the different test procedures adopted in different countries (ER/60 is equal to one when the energy transmitted to the rod during the test is equal to 60% of the potential energy of the hammer).

The chart of Figure 10 refers to an effective overburden pressure not greater than 1 kg/cm², to sandy soils with percent of fine not greater than 5, and to an earthquake having an equivalent number of representative cycles which corresponds to a magnitude M = 7.5.

For percent of fine greater than 5, an effective soil resistance index can be expressed, for practical purposes, in terms of an equivalent clean sand value by the equation:[65]

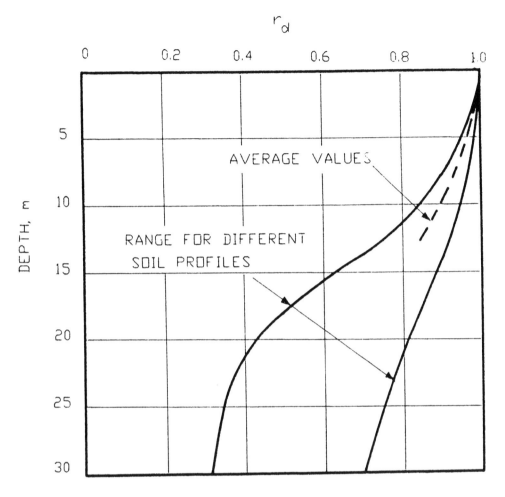

FIGURE 9. Stress reduction coefficient: proposed range of values. (From Seed, H. B. and Idriss, I. M., *J. Soil Mech. Found. Div. Am. Soc. Civil Eng.*, 97, 1249, 1971.)

$$N_{1(60)} \text{ effective } = N_{1(60)} \text{ measured } + \Delta N_{1(60)} \tag{25}$$

where $\Delta N_{1(60)}$ depends on percent of fine as shown in Table 8.

Finally, for magnitude values different from 7.5, the cyclic stress ratio causing liquefaction is determined multiplying the value, obtained from Figure 10, by the scaling factor of Table 9.

Other procedures have been proposed in order to evaluate liquefaction potential. They are based on formulations of soil constitutive laws and account for generation and dissipation of pore water pressure in evaluating both the state of shear stresses, induced by the earthquake, and the liquefaction resistance.[66-68] These procedures, however, because of modeling of soil behavior and evaluation of soil properties, are still in a research stage.

2. Settlements

Settlements of the ground surface have been observed during major earthquakes. They are determined by the following phenomena.

1. Tectonic movements (surface faulting). Movements of tectonic structures, propagating up to the surface, may cause horizontal and vertical displacements. Geologic studies

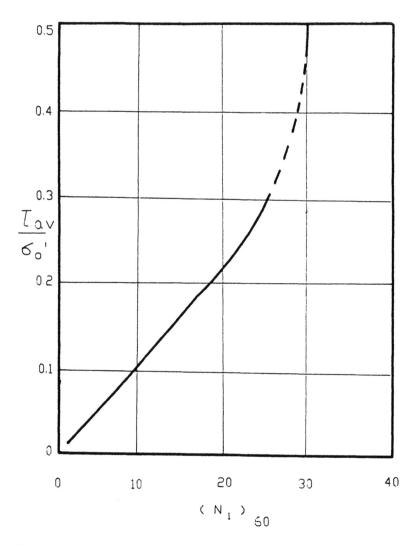

FIGURE 10.   Chart for evaluation of liquefaction resistance for sands with percent of fine not greater than 5, for an earthquake magnitude M = 7.5, and for an effective overburden pressure not greater than 1 kg/cm². (From Seed, H. B., Tokimatsu, K., Harder, L. F., and Chung, R. M., Report No. UBC/EERC-84/15, Earthquake Engineering Research Center, University of California, Berkeley, 1984.)

are required in order to investigate those faults which, during major earthquakes, may determine surface ruptures and displacements. Construction site for major hazard industrial plants must be chosen far from these faults.

2.   Compaction of unsaturated granular soils. Vibrations induced in soil deposits or earth structures (backfills), determine compaction with settlements at the surface. Settlements due to compaction phenomena can be evaluated correlating soil properties with earthquake-induced shear stresses.[69] The soil properties, as for liquefaction analysis, are determined by *in situ* tests (normally SPT) or laboratory test (cyclic triaxial or simple shear tests). The dynamic shear stresses are evaluated by simplified procedures[62] or by dynamic response analyses.[16]

3.   Dissipation of pore water pressure build-up in saturated granular soils. As just described above, dynamic shear stresses generate an increment of pore water pressure in cohesionless saturated soil deposits. Even if liquefaction does not occur, the subsequent

Table 8
SUGGESTED $\Delta N_{1(60)}$ VALUES

Fine (%)	$\Delta N_{1(60)}$
≤5	0
15	3
35	5
50	7

From Seed, H. B., Design Problems in Soil Liquefaction, Report no. UBC/EERC-86/02, Earthquake Engineering Research Center, University of California, Berkeley, 1986.

Table 9
SCALING FACTOR FOR LIQUEFACTION RESISTANCE

Earthquake magnitude	Representative cycles	Scaling factor
8.5	26	0.89
7.5	15	1.00
6.75	10	1.13
6	5—6	1.32
5.25	2—3	1.50

From Seed, H. B. and Idriss, I. M., *Ground Motions and Soil Liquefaction During Earthquakes*, Earthquake Engineering Research Institute, Berkeley, CA, 1982.

dissipation of the pore water pressure increment determines a volume reduction and settlements. Also in this case, settlements can be evaluated correlating soil properties and induced dynamic shear stresses.[69]

3. Loss of Bearing Capacity

During earthquakes building foundations apply to the underlying soil not only vertical loads but shear forces and overturning moments as well.

For surface foundations, the bearing capacity of the foundation soil is defined as the maximum vertical load that the soil can bear for a given geometric situation of the foundation structure.

In the case of dynamic-loaded foundations the eccentricity and inclination of the applied loads have to be suitably accounted for. The dynamic loads are considered as static forces and the soil-bearing capacity is evaluated by the classic methods of plasticity or considering the equilibrium of a rigid body along a given slip surface.

The soil behavior is represented by a rigid plastic model; no deformation is allowed before a given threshold value of the applied load, while an infinite deformation is considered when the threshold value is reached.

The Coulomb failure criterium is, finally, adopted, assuming that a linear relationship between the shear strength and the applied effective stress can be expressed by the soil resistance parameters: cohesion and friction angle.

Available solutions have been developed for the two-dimensional problems but it is possible to account for three-dimensional situations in an approximate way.

The method based on the theory of plasticity was initially proposed for a strip foundation, subjected to a vertical load, on a homogeneous isotropic half-space.[70] The method was, then, modified to account for different foundation shapes and for inclined and eccentric loads.[71] Caution has to be used, however, when high horizontal forces and high overturning moments are applied and with layered and inhomogeneous soil deposits. In this last case, in fact, it may be questionable to identify the soil volume, interested by failure, with a homogeneous half-space and it is preferable to apply the methods of limit equilibrium.[72-75] These methods, in fact, can account for soil layering and inhomogeneity. They, by iterative procedures, lead to the evaluation of the slip surface with the minimum factor of safety against failure.

Regarding deep foundations (pile foundations) it should be noted that, during earthquakes, piles are subjected to distortions, due to different-with-depth horizontal soil displacements. Design of pile foundations in an earthquake-resistant way should account not only for the bearing capacity of piles[76] but for resistance to differential soil displacements[77] as well.

4. Slope Stability

Slope instability could cause damage to buildings founded on slopes or located nearby.

This phenomenon may be triggered by liquefaction of sandy layers or if the induced shear stresses overcome the soil strength along a critical slip surface.

The induced shear stresses can be determined by a dynamic analysis of a plain strain finite element model of the soil slope[78] or considering pseudo-static seismic forces.[60] The comparison between the induced stresses and the soil strength is finally carried out along potential slip surfaces, determining the slip surface with the lowest factor of safety.

5. Stability of Retaining Structures

Earth pressure on retaining structures is normally determined by Mononobe-Okabe theory.[79,80] This theory modifies the classic Coulomb theory, introducing pseudo-static seismic forces and assuming that the seismic earth pressure linearly increases with depth.

A more recent theory[81] proposes the Mononobe-Okabe formulation of lateral earth pressure, recommending that the seismic increment of the earth force on the retaining structure be assumed at a point located at 60% height above the base.

III. THE GEOTECHNICAL ASPECT: SITE INVESTIGATION AND FOUNDATION STABILITY

The foundation soil is a safety-related part of the plant. It bears the static and dynamic loads applied by the structures and, in its turn, applies static and dynamic loads to the structures.

The static loads, applied by the structures are normally high and affect large areas. For this reason bearing capacity problems or high and unacceptable differential settlements between different parts of the plant can arise, especially for a soft soil condition.

The loads, transferred from the soil to the structures (as the seismic loads), may be important for structural design and systems or components qualification.

Last, but not least, it should be noted that the geotechnical characteristics of the site qualify not only the plant design but the site selection as well.

A. Geotechnical Investigations

Geotechnical investigations are required in order to know the medium (the foundation soil), which has been assigned to receive and transfer loads from and to the structures. ''To know'', in this case, means suitable description of the stratigraphic characteristics, the physical properties, and the mechanical (static and dynamic) behavior of a soil region, significant from an engineering point of view.

The investigations should be performed according to an investigation plan, whose extension and degree of detail depends both on the complexity of the geologic and geotechnical situation of the site and on the characteristics and the potential hazard of the plant to be constructed. Moreover the investigation means (boreholes, *in situ* tests, laboratory tests) should be selected according both to the soil characteristics and to the design requirements.

Site investigations are, normally, performed in different stages.

A first stage can be provided for selecting the site within a large area. In this case geotechnical information are not specifically aimed at design but at giving elements for site selection together with other environment parameters. Moreover, investigations, in this stage, shall provide all the essential information for evaluating whether geologic and geotechnical hazards can affect the site, such as subsidence of the area, surface faulting, liquefaction, collapse for karst phenomena, or instability of natural slopes or earth dams.

A second more-detailed investigation stage, planned taking into account the results of the first stage, aims at collecting all the information about stratigraphic characteristics, groundwater level, physical and mechanical (static and dynamic) properties of the foundation soil, necessary for plant design.

Geotechnical investigations have to be carried on, during the construction stage of the plant, setting up a geotechnical instrumentation, in order to collect further data about the soil behavior and review, if necessary, the design specification related to the foundation soil.

Finally, during the operation of the plant, a suitable geotechnical and seismic instrumentation of the site aims at checking the plant behavior for long-term phenomena (such as soil consolidation) or in case of earthquakes.

All the information and results, gathered during *in situ* and laboratory investigations, have to be analyzed and synthesized in a model of the foundation soil (the design soil profile) which represents the collecting point of all the data obtained during the soil investigation stages and the starting point for the design analyses.

The design soil profile should describe the best-estimate values and the variability of the geotechnical parameters to be used in design.

B. Geotechnical Analyses

Having determined the soil properties, the potential causes of instability of the plant foundations have to be analyzed and engineering solutions have to be proposed in order to overcome those problems which represent a hazard for the plant. In particular differential settlements of the foundation soil and loss of bearing capacity[82] should be carefully considered.

Settlements may have different origins: loading (loads applied by the structures and lowering of the groundwater level), unloading (excavations, interaction with retaining structures, and tunnels), and natural events (subsidence, chemical actions, erosion, variation of the hydrogeologic regime, and earthquakes).

In soil deformation it is possible to distinguish an immediate component from a long-term component.

The immediate component is that portion of settlements which occurs simultaneously with load application. It is due to a rearrangement of soil grains with no volume variation, but only with a shape variation.

The long-term or consolidation component results from the gradual dissipation of the pore water pressure, induced by the applied loads, with expulsion of water and contemporaneous compression of soil. It is a function not only of the applied loads but of time as well. The consolidation time (i.e., the time during which the consolidation settlements occur) depends on soil permeability and drainage conditions. For thick cohesive soil layers of low permeability, the consolidation time ranges from a few years to decades, while, in sandy soils, which are more permeable, the pore pressure increments, induced by dynamic stresses, dissipate in a few hours.

The allowable maximum differential settlements depend on type and function of structures, systems, and components.

Settlements, which occur during construction and before connections between the different part of the plant take place, can normally by supported by the structures, if properly foreseen, and are not important for systems and components. On the contrary, if long-term settlements are expected, it is necessary to account for differential displacements in order to design not only the building structures but the connections between the parts of the plant located on different buildings as well.

Loss of bearing capacity may have different origins: unforeseen static or dynamic loads, higher than those considered in design, or loss of soil strength because of seismic-induced stresses (liquefaction) or due to seepage phenomenon (quick condition).

For shallow foundations, loss of bearing capacity occurs with a shear failure of the foundation soil along a continuous slip surface.

For deep foundations, loss of bearing capacity may be induced by settlements of the pile group or by dynamic stresses applied to the piles during earthquakes.

Foundation stability can be improved acting on the foundation soil by stabilization techniques,[83] or on the structure, selecting the more suitable foundation type.

It is possible to reduce the consolidation time, speeding up settlements (e.g., driving vertical drains — sand drains — in cohesive soils), or to improve the static and dynamic strength of sandy soils (e.g., with a vibroflotation technique).

Foundation stability can be improved by piles, transferring the loads from the structure to deep and more-consistent soil layers. The consequence of differential settlements can be limited designing a unique mat foundation and housing on it the different connected parts of the plant.

Both soil-improvement techniques and foundation types should be selected, however, on a case-by-case basis, taking into account not only the soil characteristics (stratigraphy, physical and mechanical properties, and groundwater regime) but the plant structural and operational characteristics as well.

REFERENCES

1. **Bolt, B. A.**, *Earthquake A Primer*, W. H. Freeman, San Francisco, 1978.
2. **Hallam, A.**, *A Revolution in the Earth Sciences: From Continental Drift to Plate Tectonics*, Oxford University Press, Oxford, 1973.
3. **Reid, H. F.**, The mechanics of the earthquake, The California Earthquake of April 18 1906, Report of the State Earthquake Investigation Commission, Carnegie Institution, Washington, D.C., 1910.
4. **Richter, C. F.**, An instrumental earthquake magnitude scale, *Bull. Seismol. Soc. Am.*, 25, 1 1935.
5. **Karnik, V.**, *Seismicity of the European Area, Part 1*, D. Reidel, Dordrecht, The Netherlands, 1969, 64.
6. **Nuttli, O. W. and Zollweg, J. E.**, The relation between felt area and magnitude for central United States earthquakes, *Bull. Seismol. Soc. Am.*, 64, 73, 1974.
7. **Campbell, K. W.**, Strong motion attenuation relations: a ten-year perspective, *Earthquake Spectra*, 1, 759, 1985.
8. **Campbell, K. W.**, Near-source attenuation of peak horizontal acceleration, *Bull. Seismol. Soc. Am.*, 71, 2039, 1981.
9. **Joyner, W. B. and Boore, D. M.**, Peak horizontal acceleration and velocity from strong motion records including records from the 1979 Imperial Valley, California earthquake, *Bull. Seismol. Soc. Am.*, 71, 2011, 1981.
10. **Sabetta, F. and Pugliese, A.**, Attenuation of peak horizontal acceleration and velocity from Italian strong motion records, *Bull. Seismol. Soc. Am.*, 77, 1497, 1987.
11. **Newmark, N. M. and Hall, W. J.**, Development of Criteria for Seismic Review of Selected Nuclear Power Plants, Report NUREG/CR-0098, U.S. Nuclear Regulatory Commission, Washington, D.C., 1978.

12. **Seed, H. B. and Idriss, I. M.**, *Ground Motions and Soil Liquefaction During Earthquakes,* Earthquake Engineering Research Institute, Berkeley, CA, 1982.
13. **Seed, H. B. and Alonso, G. J. C.**, Effects of soil-structure interaction in the Caracas earthquake of 1967, Proc. 1st Venezuelan Conference on Seismology and Earthquake Engineering, 1974.
14. National University of Mexico, Effects of the September 19, 1985 Earthquake in the Buildings of Mexico City, Preliminary report, Institute of Engineering, National University of Mexico, Mexico City, 1985.
15. **Roesset, J. M.**, Soil amplification of earthquakes, in *Numerical Methods in Geotechnical Engineering,* Desai, C. S., Cristian, J. T., Eds., McGraw-Hill, New York, 1977, chap. 19.
16. **Schnabel, P. B., Lysmer, J., and Seed, H. B.**, SHAKE: A Computer Program for Earthquake Response of Horizontally Layered Sites, Report EERC-72, Earthquake Engineering Research Center, University of California, Berkeley, 1972.
17. Structural Engineers Association of California, Seismology Committee, Recommended Lateral Force Requirements and Commentary, San Francisco, 1980.
18. Applied Technology Council, Tentative Provisions for the Development of Seismic Regulations for Buildings, ATC 3-06, National Bureau of Standards, Washington, D.C., 1978.
19. **Gutenberg, B. and Richter, C. F.**, Earthquake magnitude, intensity, energy and acceleration (second paper), *Bull. Seismol. Soc. Am.,* 46, 143, 1956.
20. **Newmark, N. M. and Hall, W. J.**, *Earthquake Spectra and Design,* Earthquake Engineering Research Institute, Berkeley, 1983, 37.
21. **Campbell, K. W.**, The Effects of Site Characteristics and Near Source Recordings of Strong Ground Motion, Open-file report 83-845, U.S. Geological Survey, Proc. of Conference XXII: A Workshop on Site Specific Effects of Soil and Rock on Ground Motion and Their Implications for Earthquake-Resistant Design, Santa Fe, NM, 1983.
22. **Joyner, W. B. and Boore, D. M.**, Prediction of Earthquake Response Spectra, Open-File Report 82-977, U.S. Geological Survey, Washington, D.C., 1982.
23. U.S. Atomic Energy Commission, Design Response Spectra for Seismic Design of Nuclear Power Plants, Regulatory Guide 1.60, U.S. Atomic Energy Commission, Washington, D.C., 1973.
24. **Trifunac, M. D. and Westermo, B. D.**, Duration of strong earthquake shaking, *Soil Dynamics and Earthquake Engineering,* 1, 117, 1982.
25. **Vanmarcke, E. H. and Lai, S. P.**, Strong motion duration and RMS amplitude for earthquake records, *Bull. Seismol. Soc. Am.,* 70, 1293, 1980.
26. **Idriss, I. M.**, Evaluating seismic risk in engineering practice, in *Proc. 11th Int. Conf. Soil Mechanics and Foundation Engineering,* Vol. 1, A. A. Balkema, Rotterdam, 1985, 255.
27. **Slemmons, D. B.**, Determination of design earthquake magnitude for microzonation, Proc. 3rd Int. Earthquake Microzonation Conference, Vol. 1, Seattle, 1982.
28. **Wyss, M.**, Estimating maximum expectable magnitude of earthquake from fault dimensions, *Geology,* 7, 336, 1979.
29. **Lysmer, J., Udaka, T., Tsai, C. F., and Seed, H. B.**, FLUSH: a Computer Program for Approximate 3-D Analysis of Soil Structure Interaction Problems, Report EERC 75-30, Earthquake Engineering Research Center, University of California, Berkeley, 1979.
30. **Kausel, E. and Roesset, J. M.**, Soil-structure interaction problems for nuclear containment structure, Proc. Am. Soc. Civil Engineers, Power Division Specialty Conference, Denver, 1974.
31. **Seed, H. B. and Idriss, I. M.**, Influences of soil conditions on ground motion during earthquakes, *J. Soil Mech. Found. Div. Am. Soc. Civil Eng.,* 95, 1199, 1969.
32. **Gazetas, G.**, Analysis of machine foundation vibrations: state of the art, *Soil Dyn. Earthquake Eng.,* 2, 3, 1983.
33. **Johnson, J. J.**, Soil Structure Interaction: The Status of Current Analysis Methods and Research, Report Nureg/CR 1780, U.S. Nuclear Regulatory Commission, Washington, D.C., 1981.
34. **Richart, F. E., Jr. and Woods, R. D.**, *Vibrations of Soils and Foundations,* Prentice-Hall, Englewood Cliffs, NJ, 1970.
35. **Kausel, E. and Ushijima, R.**, Vertical and Torsional Stiffness of Cylindrical Footings, Research Report R76-6, Massachusetts Institute of Technology, Cambridge, 1979.
36. **Elsabee, F. and Morray, J. P.**, Dynamic Behavior of Embedded Foundations, Research Report, R77-33, Massachusetts Institute of Technology, Cambridge, 1977.
37. American Society of Civil Engineers, Ad Hoc Group on Soil-Structure Interaction, Analyses for Soil Structure Interaction Effects for Nuclear Power Plants, American Society of Civil Engineers, 1979.
38. **Clough, R. W. and Penzien, J.**, *Dynamics of Structures,* McGraw Hill, New York, 1975.
39. U.S. Atomic Energy Commission, Damping Values for Seismic Design of Nuclear Power Plants, Regulatory Guide 1.61, U.S. Atomic Energy Commission, Washington, D.C., 1973.
40. **Tsai, N. C.**, Modal damping for soil-structure interaction, *J. Eng. Mech. Div. Am. Soc. Civil Eng.,* 100, 323, 1974.

41. **Roesset, J. M., Whitman, R. V., and Dobry, R.,** Modal Analysis for Structures with Foundation Interaction, *Journal of the Structural Division,* Am. Soc. Civil Engineers, 99, 399, 1973.

42. **Newmark, N. M.,** A response spectrum approach for inelastic seismic design of nuclear reactor facilities, Proc. 3rd Int. Conf. Structural Mechanics in Reactor Technology, Vol. 4, part. K, London, 1975.

43. **Hadjian, A. H.,** On the decoupling of secondary systems for seismic analysis, Proc. 6th World Conference on Earthquake Engineering, New Delhi, 1977.

44. U.S. Nuclear Regulatory Commission, Development of Floor Design Response Spectra for Seismic Design of Floor Supported Equipments or Components, Regulatory Guide 1.122, U.S. Nuclear Regulatory Commission, Washington, D.C., 1978.

45. **Newmark, N. M. and Rosenblueth, E.,** *Fundamentals of Earthquake Engineering,* Prentice-Hall, Englewood Cliffs, NJ, 1971, 197.

46. U.S. Nuclear Regulatory Commission, Combining Modal Responses and Spatial Components in Seismic Response Analysis, Regulatory Guide 1.92, U.S. Nuclear Regulatory Commission, Washington, D.C., 1976.

47. **Ohsaki, Y.,** Niigata earthquakes, 1964 building damage and condition, *Soils Found. Jpn. Soc. Soil Mech. Found. Eng.,* 6, 14, 1976.

48. **Seed, H. B. and Wilson, S. D.,** The Turnagain heights lansdlide, Anchorage, Alaska, *J. Soil Mech. Found. Div. Am. Soc. Civil Eng.,* 93, 325, 1967.

49. **Ohsaki, Y.,** Effects of sand compaction on liquefaction during the Tokachi-Oki earthquake, *Soils Found. Jpn. Soc. Soil Mech. Found. Eng.,* 10, 112, 1970.

50. **Seed, H. B.,** State of the art lecture on the influence of local soil conditions on earthquake damage, Proc. 7th Int. Conf. Soil Mechanics and Foundation Engineering, Mexico City, 1969.

51. **Ishiara, K. and Koga, Y.,** Case studies of liquefaction in the 1964 Niigata earthquake, *Soils Found. Jpn. Soc. Soil Mech. Found. Eng.,* 21, 35, 1981.

52. **Tohno, I. and Yasuda, S.,** Liquefaction of the ground during the 1978 Miyagiken-Oki earthquake, *Soils Found. Jpn. Soc. Soil Mech. Found. Eng.,* 21, 18, 1981.

53. **Bennett, M. J., Youd, T. L., Harp, E. L., and Wieczorek, G. F.,** Subsurface Investigation of Liquefaction, Imperial Valley Earthquake California, October 15, 1979, Open-File Report 81-502, U.S. Geological Survey, Menlo Park, CA, 1981.

54. **Youd, T. L. and Hoose, S. N.,** Liquefaction susceptibility and geologic setting, Proc. 6th World Conference on Earthquake Engineering, New Delhi, 1977.

55. **Tsuchida, H.,** Prediction and countermeasure against the liquefaction in sand deposits, Abstr. Semin. Port of Harbor Research Institute (in Japanese), 1970.

56. **Seed, H. B., Tokimatsu, K., Harder, L. F., and Chung, R. M.,** The influence of SPT Procedures in Soil Liquefaction Resistance Evaluations, Report no. UBC/EERC-84/15, Earthquake Engineering Research Center, University of California, Berkeley, 1984.

57. **Seed, H. B. and Lee, K. L.,** Liquefaction of saturated sands during cyclic loading, *J. Soil Mech. Found. Div. Am. Soc. Civil Eng.,* 92, 105, 1966.

58. **Seed, H. B. and Peacock, W. H.,** Test procedures for measuring soil liquefaction characteristics, *J. Soil Mech. Found. Div. Am. Soc. Civil Eng.,* 97, 1099, 1971.

59. **Seed, H. B.,** Soil liquefaction and cyclic mobility evaluation for level ground during earthquakes, *J. Geotech. Eng. Div. Am. Soc. Civil Eng.,* 105, 1979.

60. **Ishiara, K.,** Stability of natural deposits during earthquakes, *Proc. 11th Conf. Soil Mechanics and Foundation Engineering,* Vol. 1, A.A. Balkema, Rotterdam, 1985, 321.

61. **Finn, W. D. L.,** Liquefaction potential: developments since 1976, Int. Conf. Recent Advances in Geotechnical Earthquake Engineering and Soil Dynamics, St. Louis, 1981.

62. **Seed, H. B. and Idriss, I. M.,** Simplified procedures for evaluation soil liquefaction potential, *J. Soil Mech. Found. Div. Am. Soc. Civil Eng.,* 97, 1249, 1971.

63. **Seed, H. B. and De Alba, P.,** Use of SPT and CPT tests for evaluating the liquefaction resistance of sands, Proc. Am. Soc. Civil Engineers Specialty Conference on Use of *In Situ* Tests in Geotechnical Engineering, Balcksburg, 1986.

64. **Liao, S. S. C. and Whitman, R. V.,** Overburden Correction Factors for SPT in Sand, *J. Geotech. Eng. Am. Soc. Civil Eng.,* 112, 373, 1986.

65. **Seed, H. B.,** Design Problems in Soil Liquefaction, Report no. UBC/EERC-86/02, Earthquake Engineering Research Center, University of California, Berkeley, 1986.

66. **Lee, M. K. W. and Finn, D. L.,** DESRA-1: Program for the Dynamic Effective Stress Response Analyses of Soil Deposit Including Liquefaction Evaluation, Soil Mechanics Series no. 36, University of British Columbia, Vancouver, 1975.

67. **Martin, P. P. and Seed, H. B.,** Simplified procedure for effective stress analysis of ground response, *J. Geotech. Eng. Div. Am. Soc. Civil Eng.,* 105, 739, 1975.

68. **Ghaboussi, J. and Dikemen, S. U.,** Effective stress analysis of seismic response and liquefaction: case studies, *J. Geotech. Eng. Div. Am. Soc. Civil Eng.,* 110, 645, 1984.

69. **Tokimatsu, K. and Seed, H. B.,** Simplified Procedures for the Evaluation of Settlements in Clean Sands, Report UCB/EERC-84/16, Earthquake Engineering Research Center, University of California, Berkeley, 1984.
70. **Terzaghi, K.,** *Theoretical Soil Mechanics,* John Wiley & Sons, New York, 1943.
71. **Vesic, A. D.,** Bearing capacity of shallow foundations, in *Foundation Engineering Handbook,* Winterkorn, H. F. and Fang, H. Y., Eds., Van Nostrand Reinhold, New York, 1975, chap. 3.
72. **Bell, J. M.,** General slope stability analysis, *J. Soil Mech. Found. Div. Am. Soc. Civil Eng.,* 94, 1253, 1968.
73. **Bishop, A. W.,** The use of the slip circle in the stability analysis of slopes, *Geotechnique,* 5, 7, 1955.
74. **Janbu, A. W.,** Earth Pressure and Bearing Capacity Calculations by Generalized Procedure of Slices, 4th Int. Conf. Soil Mechanics and Foundation Engineering, London, 1957.
75. **Morgenstern, N. R. and Price, V. E.,** The analysis of the stability of general slip surfaces, *Geotechnique,* 15, 79, 1965.
76. **Poulos, H. G. and Davis, E. H.,** *Pile Foundation Analysis and Design,* John Wiley & Sons, New York, 1980.
77. **Penzien, J.,** Soil-pile foundation interaction, in *Earthquake Engineering,* Wiegel, R. L., Eds., Prentice-Hall, Englewood Cliffs, NJ, 1970, chap. 14.
78. **Idriss, I. M., Lysmer, J., Hwang, R., and Seed, H. B.,** QUAD-4: A Computer Program for Evaluating the Seismic Response of Soil Structures by Variable Damping Finite Element Procedure, Report EERC-73/6, Earthquake Engineering Research Center, University of California, Berkeley, 1973.
79. **Okabe, S.,** General theory of earth pressure, *J. Jpn. Soc. Civil Eng.,* 12, 1926.
80. **Mononobe, N. and Matsuo, M.,** On the determination of earth pressure during earthquake, Proc. World Engineering Congress, Tokyo, 1929.
81. **Seed, H. B. and Whitman, R. V.,** Design of earth retaining structures for dynamic loads, Proc. Am. Soc. Civil Engineers, Specialty Conference on Lateral Stresses in the Ground and Design of Earth-Retaining Structures, Ithaca, 1970.
82. **Terzaghi, K. and Peck, R. B.,** *Soil Mechanics in Engineering Practice,* 2nd ed., John Wiley & Sons, New York, 1967.
83. **Mitchell, J. K.,** Soil improvement — state-of-art report, *Proc. 10th Int. Conf. Soil Mechanics and Foundation Engineering,* Vol. 4, A.A. Balkema, Rotterdam, 1981, 509.

Chapter 2

NATURAL EVENTS: SEVERE METEOROLOGICAL EVENTS

Alberto Ferreli

TABLE OF CONTENTS

I. INTRODUCTION*

Tornadoes, tropical cyclones, and floods are considered among the most severe external events which may affect the safety of a plant. The extent of protection required against these events depends on their probability of occurrence and on the hazard the plant may present under their effects.

Regarding nuclear power plants (NPPs), it is required that "safety-related"** structures, systems, and components be protected in order to carry out their safety function under the most severe natural events and in absence of all the external electrical supplies.

In the evaluation of the "probable maximum event" to be used as a basis for the design of the plant, two approaches are followed: the deterministic approach, mainly developed in the U.S., and the stochastic one, which is mainly applied in the European countries.

In the deterministic approach the reference maximum event is derived from a comprehensive application of the most severe, reasonably possible meteorological and hydrologic factors. Historical data may be used as input to mathematical models representing physical relationships among different parameters of the actual site situation.

The evaluation leads normally to conservative results; sufficient margins are taken in relation to the accuracy and the quantity of the historical data and the period of time in which they have been collected.

The stochastic approach is used when sufficient records are available over a sufficient period of time; it is applied, for instance, in some European countries where meteorological and hydrologic records have existed since the beginning of the century, to determine the probable maximum flood along rivers and coastal sites. In the analysis, climatic and hydrometeorological changes must be considered, as well as new dam constructions and river deviations for site along streams, and changes in the shoreline geometry for coastal sites.

II. TORNADOES

Tornadoes are violent cyclonic storms, relatively small in diameter (a few hundred meters), but with very rapidly rotating winds, whose speed can reach, at the edges, about 500 km/ h. They are characterized near the center by a sudden pressure drop which can attain values up to 20 kPa (0.2 bar). For the violence of their winds, tornadoes present a greater hazard for persons and properties than other types of storms. Moreover, the intense updraft due to the vortex is capable of lifting into air quite heavy objects, such as trees and cars, which may become projectiles hurled against buildings.

Tornadoes are most frequent in America (average frequency: 600 to 1000 per year per whole country), whereas they are much less frequent and severe in Europe (average frequency 1 to 5 per year per country).

A. Tornado Characteristics

Because of inherent difficulty in directly observing tornadoes — time and location of its occurrence cannot be predicted with sufficient accuracy — their properties have to be defined indirectly by observation of their effects. The characteristics which are relevant for design purposes are

---

* The considerations contained in this chapter are mainly derived from the experience gained in the nuclear applications. Their extrapolation to other "major hazard" activities should be made carefully, taking into account the effective potential of hazard these activities may represent under the effect of the events considered.

** In nuclear activities the following items are defined as safety-related: structures, systems, and components which are part of the reactor pressure boundary or are necessary to shut down the reactor, to keep it in a shutdown condition, to remove the decay heat from the core and the containment, and to avoid uncontrolled release of radioactive material.

- Rotational wind speed
- Translational wind speed
- Radius of maximum rotational wind speed
- Pressure drop
- Rate of pressure drop

The value of these parameters are then used to characterize tornado loads on structures and buildings of the plant.

With specific reference to NPPs, in U.S. design basis tornadoes are defined for each of the three tornado intensity regions within its territory. In Table 1,[4] the tornado characteristics for these three zones are listed.

B. Tornado Missile

Provisions for tornado protection should also consider the effects of loads caused by tornado-generated missiles, such as steel pipes, wooden planks, and cars, which should be characterized according to their weight, speed, maximum height, and area of impact.

With reference to the U.S. regulations for NPPs, the characteristics of the missiles impelled by the most violent of the tornadoes above described are listed in Table 2.[15]

C. Plant Protection Against Tornadoes

In order to provide adequate protection against tornado, safety-related systems and components are identified and protected into structures designed to withstand the effect of the design basis tornado.

Normally tornado loads are combined with normal operating stresses and, because of the low probability of the event, they are not combined with other extreme loading conditions deriving from internal accidents or external events, such as earthquakes or floods.

Criteria for transforming tornado wind velocities into effective pressures applied to structures are delineated in References 28 and 29.

Global effects upon structures shall be evaluated combining tornado pressure loads with other loads, according to national regulations. Three-dimensional static models and finite elements analysis are generally used, simplified models being employed only in specific cases.

Local effects associated with the impact of missiles are outlined in Figure 1 for structures in reinforced concrete.

For low-speed values the structure behaves almost elastically, and the missile ricochets backwards.

As the speed increases fragments are projected in the opposite direction to the missile (spalling), and a crater is formed around the area of impact (Figure 1A).

For greater values of the speed, splinters are projected from the surface opposite to the impact (scabbing), and further increases in speed produce perforation of the wall (Figures 1B and C). Scabbed layers, projected from the surface, may also represent a threat to any surrounding component (Figure 1D).

Sufficient thickness of concrete should be provided to prevent perforation, spalling, or scabbing in the event of a missile impact. Scabbing effects need not be considered if a suitable liner is present on the inside of the structures.

Several empirical equations are available[30] to estimate missile penetration into concrete and to determine the required barrier thickness.

In Table 3 minimum barrier thicknesses for protection against tornado-generated missiles for the three tornado regions presented in Table 1, are reported.

Missile penetration tests into steel plates have been conducted in the U.S. by the Stanford

Table 1
TORNADO CHARACTERISTICS

Region	Maximum wind speed[a] (km/h)	Rotation speed (km/h)	Maximum translational speed (km/h)	Radius of maximum rotational speed (km/h)	Pressure drop (10^4 Pa)	Rate of pressure drop (10^4 Pa/s)
I	576	467	109	50	2.1	1.4
II	480	384	96	50	1.6	0.8
III	386	306	80	50	1.0	0.4

[a] The maximum wind speed is the sum of the rotational speed component and the maximum translational speed component.

Table 2
TORNADO MISSILE SPECTRUM

Missile	Length (m)	Mass (kg)	Velocity[a] (m/s)
Wood plank, 9.2 cm × 28.9 cm	3.6	52	83
Steel rod, diameter 2.54 cm	0.9	4	51
Steel pipe, diameter 16.8 cm, schedule 40	4.5	130	52
Steel pipe, diameter 32 cm, schedule 40	4.5	340	47
Wooden telegraph post, diameter 34.3 cm	10.6	510	55
Automobile, (frontal area: 1.3m × 2.0 m)	5.0	1810	59

[a] Velocities are horizontal velocities. For vertical velocities, 70% of the horizontal velocities is acceptable.

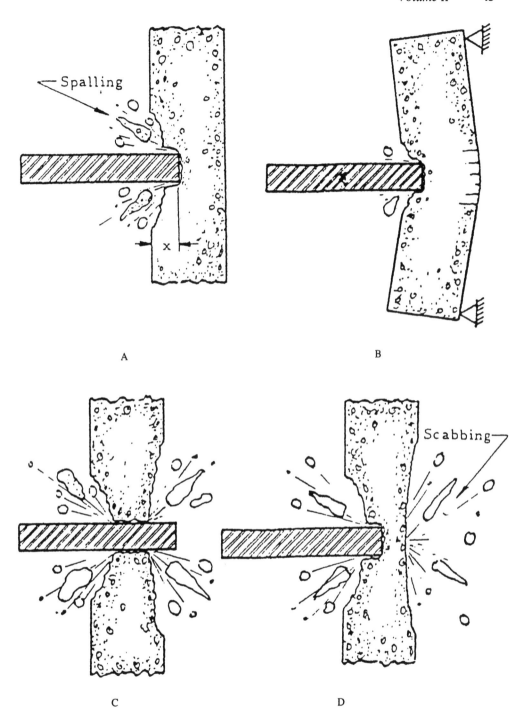

FIGURE 1. (A) Missile penetration and spalling, (B) overall target response, (C) perforation, and (D) target scabbing.

Table 3
MINIMUM ACCEPTABLE BARRIER THICKNESS
REQUIREMENTS FOR LOCAL DAMAGE PREDICTION
AGAINST TORNADO-GENERATED MISSILES

	Concrete strength (N/cm²)	Wall thickness (cm)	Roof thickness (cm)
Region I	2000	56	44
	2750	50	40
	3500	44	34
Region II	2000	40	32
	2750	34	27
	3500	32	24
Region III	2000	15	15
	2750	15	15
	3500	15	15

Research Institute;[31] other empirical correlations have been derived by the Ballistic Research Laboratory.[32]

Further information on barrier design may be found in Section V.

III. TROPICAL CYCLONES

Tropical cyclones are violent cyclonic storms, very large in diameter (100 km or more), characterized by a vast rotating mass of warm, humid air, and a marked pressure drop from the periphery to the center, which may attain 90 kPa (0.9 bar).

The physical phenomena which lead to a cyclone formation are very complex; the source energy is provided by the warm sea in the form of water vapor, which releases its latent heat when it condenses and forms rain. For this reason they are generally formed on the surface of the sea and represent a concern for coastal sites in the tropical regions; they are frequent in the Bay of Bengal, the Arabian Sea, and the southwest Indian Ocean, while in the Atlantic Ocean they are known as "hurricanes" and in the western North Pacific as "typhoons".

The core of the cyclone, called the "eye", is characterized by light winds and lightly clouded sky — the winds increase roughly near the outer edge of the eye — where they may reach 360 km/h, and then they decrease gradually with distance.

Wind speed depends on different factors, such as pressure drop, radius of the maximum winds, translational speed, etc. Also associated with cyclones are heavy rains and surges, which may cause floodings.

For design purposes, a "probable maximum tropical cyclone" (PMTC) is derived from a combination of metereological parameters giving the highest persistent wind speed which can occur at a specified site. Procedures for evaluating the PMTC are contained in Reference 23.

IV. FLOODS

Floods are defined as temporally limited swelling of the surface of a water body beyond a certain specified level. Uncontrolled floods may cause considerable damages on human well-being.

For sites on rivers they result mainly from excessive rainfall over brief periods of time.

For sites along the coasts, potentially disastrous floods may arise from storm tides and tsunamis.*

A. Design Basis Flood

Design basis floods are commonly evaluated by extrapolating existing records and data up to probabilities regarded still of interest for plant protection; all relevant events should be considered and an analysis should be performed to evaluate which events may occur simultaneously.

For sites along streams the design basis flood should be derived both from the characteristics of the drainage area and from a combination of unfavorable hydrometereological conditions — such as probable maximum precipitation, sequential storms, and snow melts — which may lead to a maximum flood runoff.

In addition to floods produced by severe hydrometeorological conditions, seismically induced floods produced by dam failures or landslides should be considered.

On sea coasts and estuaries, combination of unfavorable circumstances such as tide and wind conditions and atmospheric low pressure should be taken into account, if their occurrence probability is not low enough.

As an example of the practices used in the nuclear applications, in France, for sites on rivers, the maximum design flood (MDF) is assumed as the maximum of the following:[35]

1. The level reached from a flow 15% greater than that estimated for the 1000-year** flood

2. Maximum historical flood — or the 100-year flood, if more severe — combined with most penalizing dam failure, and average tide

For coastal sites, the MDF corresponds to the combination of the maximum calculated tide and the 1000-year high sea level. The introduction of a safety margin is required for Mediterranean coastal sites.

For estuary sites, the MDF corresponds to the most severe of the following combinations of events:

1. 1000 year flood and maximum calculated tide
2. Maximum historical flood — or the 100-year flood, if more severe — penalizing dam failure, and average tide
3. 1000-year sea level and maximum calculated tide

B. Plant Protection Against Floods

All safety-related components shall be protected such that they can fulfill their safety functions under the most severe load conditions derived from the design basis flood. Main permanent provisions against flood should include, in combination or in alternative:

---

* Tsunamis are mountainous sea waves caused by vertical displacements of the sea bottom associated with earthquakes. They are very frequent and severe on the Pacific coasts of the U.S. and Japan. Their occurrence probability is low in the Atlantic coasts of Europe, where no damaging tsunamis have been recorded in the past. Since the seismic character of the Mediterranean area, they may occur in the Greek and Italian coasts, where in 1908 an earthquakes-generated tsunami of about 8 m destroyed Messina.

** 1000-year flood is a flood which has an occurrence probability of once every 1000 years.

1.   Choice of a "dry site"* under the most severe event or combination of events
2.   Elevated arrangement of the parts of the plant to be protected
3.   Elevated arrangement of the entrances and openings
4.   Watertight construction of penetrations which may fall under the expected maximum water level
5.   Watertight construction of underground facilities containing safety-related equipment to prevent the penetration of groundwater.

Accessibility for the supply of the necessary operating materials should also be ensured during the design basis flood. In combination with or in alternative to permanent provisions, temporary flood protections, such as mobile barriers or bilge pumps, may be used.

The flood protection should be supported by organizational and administrative procedures, which should ensure adequate training of the personnel, emergency procedures, and operability of all the equipment required for flood protection.

V. OTHER NATURAL EVENTS

Besides tornadoes, cyclones, and floods, other natural phenomena which may have safety implications are

1.   Extreme air and cooling water temperature
2.   Effects of snow, ice, and hail
3.   Sandstorms
4.   Lightning
5.   Possible effects of living organisms, such as impact of birds, blockage of circulating water pump intake strainers by fish or marine algae, etc.

Since the safety relevance of these phenomena varies considerably in relation to the site considered, the type and the extent of the protection should be determined on a case-by-case basis.

In particular, special attention has been received in some countries regarding the protection of the safety functions of a nuclear power plant from the potential effects of lightning, by prescribing a minimum distance between electrical circuits and the external walls of the buildings.

REFERENCES

1. Protection of Nuclear Power Plants against High Water, KTA-2207, Gesellschaft für Reaktorsicherheit, Germany, 1982.
2. Design Basis Floods for Nuclear Power Plants, Regulatory Guide 1.59, Rev. 2, U.S. Nuclear Regulatory Commission, Washington, D.C., August 1977.
3. Standard Format and Content of Safety Analysis Reports for Nuclear Power Plants. LWR Edition, Regulatory Guide 1.70, Rev. 3, U.S. Nuclear Regulatory Commission, Washington, D.C., November 1978.
4. Design Basis Tornado for Nuclear Power Plants, Regulatory Guide 1.76, U.S. Nuclear Regulatory Commission, Washington, D.C., April 1974.
5. Flood Protection for Nuclear Power Plants, Regulatory Guide 1.102, Rev. 1, U.S. Nuclear Regulatory Commission, Washington, D.C., September 1976.

* A "flood-dry site" is one where safety-related structures are so high above potential flood sources that safety from flooding is obvious or can be documented with minimum analysis.

6. Protection Against Low-Trajectory Turbine Missiles, Regulatory Guide 1.115, Rev. 1, U.S. Nuclear Regulatory Commission, Washington, D.C., July 1977.

7. Tornado Design Classification, Regulatory Guide 1.117, Rev. 1, U.S. Nuclear Regulatory Commission, Washington, D.C., April 1978.

8. Physical Models for Design and Operation of Hydraulic Structures and Systems for Nuclear Power Plants, Regulatory Guide 1.125, Rev. 1, U.S. Nuclear Regulatory Commission, Washington, D.C., October 1978.

9. Inspection of Water-Control Structures Associated with NPPs, Regulatory Guide 1.127, Rev. 1, U.S. Nuclear Regulatory Commission, Washington, D.C., March 1978.

10. Design Basis Floods for Fuel Reprocessing Plants and for Plutonium Processing and Fuel Fabricating Plants, Regulatory Guide 3.40, Rev. 1, U.S. Nuclear Regulatory Commission, Washington, D.C., December 1977.

11. Wind Loadings, Standard Review Plan 3.3.1, Rev. 2, U.S. Nuclear Regulatory Commission, Washington, D.C., July 1981.

12. Wind Loadings, Standard Review Plan 3.3.2, Rev. 2, U.S. Nuclear Regulatory Commission, Washington, D.C., July 1981.

13. Flood Protection, Standard Review Plan 3.4.1, Rev. 2, U.S. Nuclear Regulatory Commission, Washington, D.C., July 1981.

14. Analysis Procedures, Standard Review Plan 3.4.2, Rev. 2, U.S. Nuclear Regulatory Commission, Washington, D.C. July 1981.

15. Missiles Generated by Natural Phenomena, Standard Review Plan 3.5.1.4, Rev. 2, U.S. Nuclear Regulatory Commission, Washington, D. C., July 1981.

16. Site Proximity Missiles (Except Aircraft), Standard Review Plan 3.5.1.5, Rev. 1, U.S. Nuclear Regulatory Commission, Washington, D.C., July 1981.

17. Structures, Systems, and Components to be Protected from Externally Generated Missiles, Standard Review Plan 3.5.2, Rev. 2, U.S. Nuclear Regulatory Commission, Washington, D.C., July 1981.

18. Barrier Design Procedures, Standard Review Plan 3.5.3, Rev. 1, U.S. Nuclear Regulatory Commission, Washington, D.C., July 1981.

19. Safety in Nuclear Power Plant Siting. A Code of Practice, Safety Series No. 50-C-S, International Atomic Energy Agency, Vienna, 1978.

20. Design Basis Flood for NPP on River Sites, Safety Series No. 50-SG-S10A, International Atomic Energy Agency, Vienna, 1983.

21. Design Basis Flood for NPP on Coastal Sites, Safety Series No. 50-SG-S10B, International Atomic Energy Agency, Vienna, 1983.

22. Extreme Meteorological Events in NPP Siting, Excluding Tropical Cyclones, Safety Series No. 50-SG-S11A, International Atomic Energy Agency, Vienna, 1981.

23. Design Basis Tropical Cyclone for NPP, Safety Series No. 50-SG-S11B, International Atomic Energy Agency, Vienna, 1984.

24. Methodology for Coping with Accidents of External and Internal Origin in PWR Power Stations — A Comparison of the Rules and Codes of Practice in Use in Belgium, France, the Federal Republic of Germany, the United Kingdom and the United States of America, EUR 10782 EN, Commission of the European Communities, Luxembourg, 1986.

25. Prise en Compte du Risque d'Inondation d'Origine Externe, Règle Fondamentale de Sûreté I.2.e., - Ministère de l'Industrie et de la Recherche, Paris, April 1984.

26. Standard for Estimating Tornado and Extreme Wind Characteristics at Nuclear Power Sites, ANSI/ANS 2.3.1983, American Nuclear Society, 1983.

27. Determining Design Basis Flooding at Power Reactor Sites, ANSI/ANS 2.8.1981, American Nuclear Society, 1981.

28. Building Code Requirements for Minimum Design Loads in Buildings and Other Structures, ANSI A58.1, Committee A58.1, American National Standards Institute, 1982.

29. Wind Forces on Structures, ASCE Paper No. 3269, Trans. Am. Soc. Civil Engineers, Vol. 126, Part II, 1961.

30. **Kennedy, R. P.**, *A Review of Procedures for the Analysis and Design of Concrete Structures to Resist Missile Impact Effects*, Holmes and Narver, Inc., Orange, CA, 1975.

31. **Cottrell, W. B. and Savolainen, A. W.**, U.S. Reactor Containment Technology, ORNL-NSIC-5, Vol. 1, Oak Ridge National Laboratory, Oak Ridge, TN, 1965, Chap. 6.

32. **Russell, C. R.**, *Reactor Safeguards*, MacMillan, New York, 1962.

33. **Rech, R. F. and Ipson, T. W.**, Ballistic perforation dynamics, *J. Appl. Mech. Trans. ASME* 30(3), 1963.

34. **Williamson, R. A. and Alvy, R. R.**, *Impact Effect of Fragments Striking Structural Elements*, Holmes and Narver, Inc., Orange, CA, 1973.

35. Prise en Compte du Risque d'Inondation d'Origine externe — Regle Fondamentale de Sureté de la Republique Francaise, No. 1.2.e, Paris, 1984.

Chapter 3

ARTIFICIAL EVENTS: FIRE RISK

Francesco Mazzini

TABLE OF CONTENTS

I. INTRODUCTION

In large industrial plants fire can cause disastrous events not only in relation to the thermic energy developed, but also to the damage that it can do to installations and deposits, and can be the first event of a sequence of incidents whose consequences can be put into three categories:

1. Thermal radiation
2. Release of dangerous materials
3. Explosions

As a consequence, protection against fire in large industrial plants must consider a detailed risk analysis which shows all the barriers predisposed against the development of the single incident and the relating protection which their reliability is based on; only in this way is it possible to prevent fire from damaging the predisposed safety device and to establish a program of action and of fire criteria that begins from the planning and must be followed during the whole plant life.

The first problem to tackle in planning is to determine the "fire areas", that is, those parts of the plant in which it is possible to realize such conditions that the consequences of a fire, which has broken out inside them, can be kept within the area itself.

When the condition is not realized by the intrinsic safety of the plant, in relation to the small quantity of energy released, it must be realized by the right arrangement of the components and of the installations and by the right boundary conditions in order to avoid the consequences of a fire spreading outside the industrial complex and damage propagating in different parts of the plant.

The choice of the fire areas therefore conditions the layout, so as to make the problem of fire defense one of the main points of safety from the beginning of the design and it cannot be solved by the simple adoption of fire-fighting and extinguishing systems.

After having defined the fire areas on the basis of the above-mentioned condition, such condition must be verified by risk analysis in order to establish if the fire areas have been chosen in the right way, or in order to render them appropriate by using the right means of protection.

Fire protections inside the areas is realized through the following three plans of action:

1. Prevention
2. Passive protection (fire loading consistent with the fire-resistance rating of the structures, suitable fire reaction properties of the materials used, safety distances)
3. Active protection (installation of automatic fire detectors and extinguishing systems, arrangement of smoke vents and of every other device able to aid the evacuation of people and the entrance of the fire brigade)

Finally, in order to avoid the fire and its effects spreading out of the fire areas, appropriate boundary conditions must be realized.

The text represents only a guide for the designer who will recall the international standards and the technical publications which will be useful to complete and adapt his preparation for a final design.

II. FIRE AREAS

The specification of fire areas must be carried out during the layout phase after having located the centers of danger for the following three effects: (1) thermal radiation, (2) release of dangerous substances, and (3) explosions.

A. Thermal Radiation

In order to recognize fire areas connected to accidents that have as a consequence dangerous thermal radiation with possible fire propagation at a distance, one must, first of all, determine the geometry of the radiant body and then effect the intensity map of thermal radiation around and individualize the insointensity curves (expressed in watts per square centimeter) below which it is not possible that the fuels present can catch fire.

The fire area is the zone included in these curves. The distribution of values of the intensity of thermal radiation in the space surrounding of the burning mass depends on the temperature of the flame, its shape, and the geometrical characteristics of the radiant body.

A few typical cases are

1. Fire of inflammable liquids contained in tanks
2. Fire of liquid pools on the ground (e.g., the case of inflammable liquid in containment basins)
3. Fire of gas flowing out from a vessel and a pipeline
4. Fire of steam in expansion (fire balls)

In these cases it is possible to define a method of calculation to determine the geometrical characteristics of the radiant source: height of the flame transversal dimensions, diameter of the sphere in the case of a fireball, inclinations and movements caused by wind.[1-5,8-15,19]

When the radiant source is defined it is possible to go on and calculate the intensity of the thermal radiation around, according to models based on the Stefan Boltzman law, and also taking into account the presence of smoke on the flame surface and the atmospheric transmission coefficient.

So the intensity of the thermal radiation (watts per square centimeters) can be fixed at different distances and for different heights, defining the map of the risk.[6,7,16-18]

The intensity radiation value which determines the border depends on the thermic exposure necessary for the fuel to catch fire.

B. Release of Dangerous Substances

The contamination of the environment due to the release of dangerous substances may be caused by a fire for three reasons:

1. By production of dangerous substances during the combustion of the fuels present
2. By increase of temperature in the containers or the equipment contained in them
3. By control systems and the safeguards predisposed to this purpose being out of order

In all these cases it is possible to forecast concentrations of the substances released in the environment surrounding the plant, by following the laws of atmospherical diffusion.[20-22]

In the first two cases, the protection must be obtained by avoiding the possibility of a fire which may cause a quantity of energy capable of producing damage to the components starting from a detailed and precautionary analysis of fire and explosion risks.

In the third case the way of choosing the fire area changes according to the kind of plant.

For example, we can show the criteria adopted in light water nuclear reactors because this is a complex case that includes situations which may be verified in other plants too.

In nuclear plants fire presents serious risks because it can bring about the loss of components that are essential to safety and in particular those components that are necessary to shut down the core reactor and to cool it after the shutdown.

In this case the partial or total fusion of the reactor core could be possible with disastrous consequences for the surrounding territory (for hundreds of kilometers even).

In the core of the reactor we find heat sources which are products of fission formed during operation. Therefore, we need to protract the time of cooling for a while after the shutdown.

In order to drain this residual heat in light water reactors redundant cooling systems at high or low pressure are necessary, so that shutdown can always be guaranteed because the efficiency of one or more systems for the reactor cooling is guaranteed too.

In this case the fire area is as large a part as possible of the central building so that the reactor can be shut down and kept in a safe condition.

Analysis of breakdown allows us to recognize, for each component that is out of order, the other components strictly necessary to the reactor shutdown and the keeping of it in a condition of a safe shutdown.

By this analysis it is possible to group in fire areas different components or component systems that can, independently from the presence of the others, carry out the expected functions.

This solution is possible when the function carried out by a system can be guaranteed by components that have the same function in a different fire area.

We must observe that the independence of a component is not obtained by isolating the component alone but all that is necessary for its working, e.g., for those components with an electrical function, feeding systems, and measurement and control systems; therefore, the fire area must include the whole building zone where the relative electrical cables are developed.

It is important to observe that in this type of fire areas the safety of the plant is not connected to the direct protection of the components; on the contrary the areas could be completely destroyed.

It is important in this case that nuclear safety must always be guaranteed.

C. Explosions

To avoid the initiation of explosive substances caused by fire it is necessary that combustible substances should not appear in the fire area.

In the case of inflammable mixtures we should distinguish the unconfined cloud explosions from the explosion that occurs in industrial equipment or a building.

For the first, the problem of the choice of the fire areas is similar to that for the release of dangerous substances; for the second, instead, the fire areas must remain confined to the building and in the equipment itself and this condition must be respected by explosion-venting surfaces of appropriate dimension and orientation[24-26](NFPA 68).

III. PREVENTION OF FIRE

The causes of fire in industrial activities according to statistical observations are basically of an electrical nature (arcs, overheating of circuit wires, of electric motors, lightning, and static electricity.)

Other less frequent cases, always in statistical order are cigarette ends, or matches, selfcombustion, sparks, breakdown in equipment and burners in heating systems, overheating of engines and machines, explosions, and bursts.

To reduce frequent fires therefore suitable rules of behavior must be adopted and the existent technical norms must be respected, especially those referring to electrical installations.

Particular attention must be paid in places where there is danger of explosion of inflammable mixtures.

Two different events absolutely independent of each other determine the explosion of a fire of inflammable mixture, specifically:

1. The presence in the environment of a dangerous substance outside of an appropriate system of containment
2. The formation of a thermic phenomenon (arc, spark, temperature) higher than that of

the autoignition of the substance at a point in which it may come into contact with the dangerous substance

The possibility that combustible, aeriform, or powder substances, accidentally introduced into the environment, can generate a formation of explosive mixtures depends on several parameters.

The most meaningful of these are the physical state of the substance, its specific gravity, its coefficient of diffusion, and its temperature related to the conditions of the atmosphere of the environment that determine the characteristics of the convective flows.

Ventilation can act against the formation of these mixtures, either by the remixing effect or the regeneration effect of the atmosphere.

In addition to ventilation it is necessary to reduce the probability of the thermic phenomenon of initiation.

It is necessary to avoid dangerous heating and the production of sparks in machines, as well as the formation of high electrical potentials caused by static electricity.

The electric installations that are the most frequent source of arcs and overheating must respect particularly strict technical norms in order to prevent the electrical equipment itself from setting fire or causing explosion.

In this connection, we mention the NFPA 70 1984 edition of the national electrical code. In the places where the danger of fire or explosion is possible the norms propose the adoption of different kinds of electrical systems with different characteristics according to the various danger levels that we can find in those environments where the electrical systems must be installed.

The level of danger of the places depends on the nature and on the quantity of combustible substances (liquid, solid, or gaseous) present and it is very different from one zone to another of the same place, depending on many factors such as the physical state of the substances, the possibility of their introduction into the environment outside the containment system, and the conditions of ventilation.

When choosing the installation one must aim at the same level of safety for different cases so that it is uniform in every case.[27,28]

The overcurrent protections must use devices correctly rated and set.

Other factors must be taken into account during the design phase in order to cut down the causes of fires; building materials (including those used for acoustic and thermal insulation) and furnitures must be noncombustible. If this is not possible they must have good characteristics of fire reaction.

Air conditioning must use materials that must not become a source of fire in any way.

Piping for adduction of lubricating oils or of other combustible liquids must be placed so that eventual losses do not come into contact with overheated surfaces.

IV. LIMITATION OF EFFECTS IN THE FIRE AREAS

A. Passive Protection
1. Fire Resistance

The fire resistance of an element for building construction (a component of the global structure) is represented by its attitude to preserve mechanical stability and tight and thermic insulation during the fire.

The characteristic quantity that measures fire resistance is the time during which the element is exposed to the fire without loss of any of the above-said requirements.[29,30]

Obviously the fire resistance of the structures, as has been defined, cannot be referred to real fire which varies in relation to many factors because it must be an intrinsic characteristic of the structure or of the component.

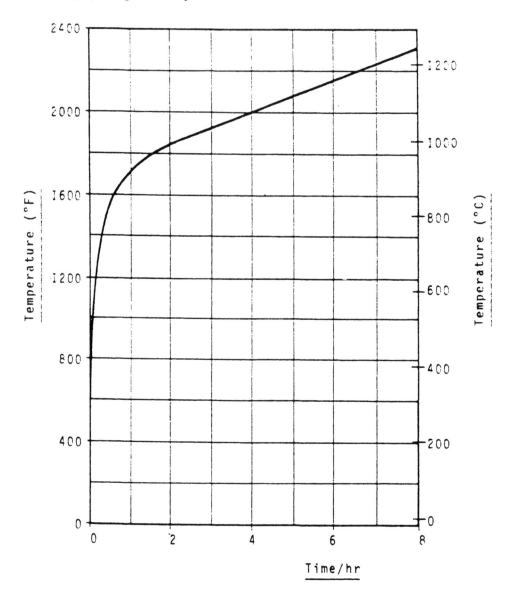

FIGURE 1.   Time-temperature standard curve.

In practice we refer to conventional fire resistance, that is, valued with a standard fire defined with a standard temperature-time curve (Figure 1) obtained with a special furnace (see NFPA 251).

The fire resistance rating of a given element is the time, in minutes, before which the element exposed to the standard temperature-time curve shows one of these three following cases:

1.   Loss of stability characteristic
2.   Transmission or production of gases and smokes from the unexposed face
3.   The temperature limit on the unexposed face reached

For the load-bearing elements the test must be carried out in true constraint and load conditions (Figures 2 to 5).

FIGURE 2.   Column HEB 180, height 3.5 m, covered with insulating varnish of 800-μm
thickness, placed under a 40-ton load, after the exposition at the standard fire in the furnace.
Fire resistance measured: 32 min.

The three requirements can all be requested or only some of them, depending on the
function that the element has in the structure.

If, for example, a separation element (door, wall, partition, floor, etc.) requires a fire-
resistance rating of 60 min, we mean that element exposed to the fire for 60 min has not
to present variations of mechanical stability, tight defects, and on its unexposed face its
temperature must not reach a certain limited value (generally 150°C).

For an element for which the essential feature is stability and not holding and thermic
insulation (beams, columns), a fire resistance rating of 60 min means that, if it is exposed
to the fire for 60 min, it must maintain only its resistance to the static charge for a time of
60 min, because it has only a structural function.

According to the fire-resistance rating, NFPA 220 standard defines the criteria for the
classification of types of building construction.

FIGURE 3.   A prestressed concrete free bearing floor, placed under the maximum moment of 5347 kg × m. Fire resistance measured: 40 min.

FIGURE 4. Intrados of the floor after the test of prestressed concrete free-bearing floor shown in Figure 3.

FIGURE 5. Particulars of the loading devices before the test.

The conventional fire resistance is the essential characteristic to classify the structural elements, but is not sufficient to define its behavior in a true fire that surely induces thermal stresses different from those of the standard temperature-time curve.

The true behavior of a fire is defined by the thermic balance between the heat power developed in the combustion, and the heat taken away in the time unit.

The heat power is due to the calorific power of the materials on fire and the combustion velocity; the latter depends on the combustion temperature, the exposed surface, the ventilation of the room, and the chemical kind of fuel.

Heat subtraction is due to smoke and combustion products (60 to 80%), to thermal radiation, to cooling by burning air which absorbs heat to reach the combustion temperature, and to the vaporization of the dampness contained in the fuel that also produces a slowing down of combustion.

Other factors which influence the development of a true fire are the geometry of the room (principally the ratio of the height with the plane surface), the features of heat transmission of the external walls, the size and the position of the windows, and more generally of the apertures. In Figure 6 the behavior of a true fire burning wood in a room of given dimensions is reported.

We can observe that the fire presents three fundamental phases; the first is the ignition phase, in which the temperature increases slowly because the subtracted heat is considerable, mostly for the presence of dampness in the fuel.

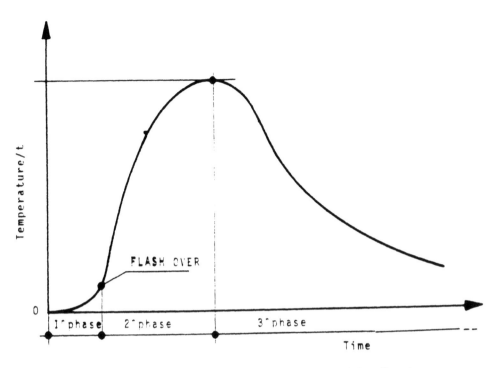

FIGURE 6.   Behavior of a real fire burning wood for a room of given dimensions.

The ignition phase is followed by that of the full fire, the beginning point of this phase is called the flash over. At this point we have a large increase in the temperature-time derivative.

After this point the thermal balance is in favor of the developed heat and so the temperature begins to increase rapidly.

With the increase in the temperature there is an increase in the radiation until the thermal equilibrium reaches the maximum point.

The third phase is that of the cooling and extinguishment. The ventilation increases the combustion velocity for the continuous flux of new comburent; there is however also a cooling effect because the new fresh air subtracts heat to reach the combustion temperature. Taking into account this phenomenon, the coefficient to introduce into the calculations to take into account the ventilation is the "ventilation factor", Fv:

$$Fv = Af/At \times H$$

where Af is the area of the windows, At is the total area of the walls plus the area of the ceiling and of the floor, and H is the arithmetic average of the vertical dimensions of the ventilation apertures.

The ventilation factor (Fv) allows us to calculate the real behavior of the fire in the different conditions independently from the geometrical characteristics of the room.

The curves in Figure 7 have been calculated for a room with a rectangular plane whose sides are 5 m × 7.5 m and a height of 4 m for different ventilation conditions, considering standard wood as fuel.

The method used is reported in the official Italian National Research Committee bulletin (technical regulations n VII n 37).[31]

The ventilation influences the temperature and the duration of the fire.

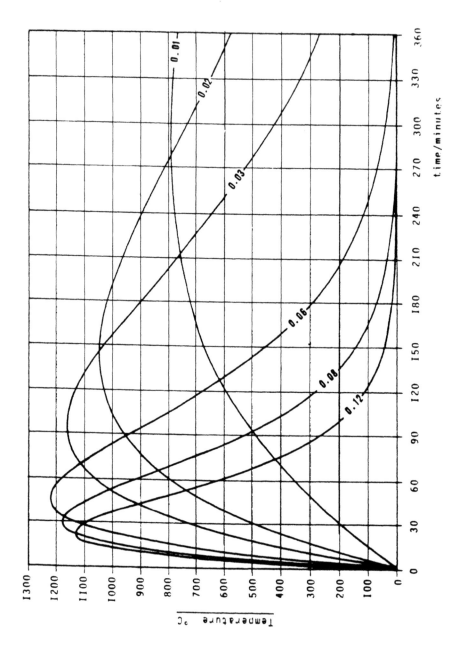

FIGURE 7. Behavior of a real fire burning wood for a room of 5 m × 7.5 m sides and 4-m height and for different ventilation conditions.

The maximum temperature reached during the fire increases by small Fv values; after a maximum it decreases.

The duration of the fire always decreases with an increasing ventilation.

The building materials are characterized by the fire resistance to the standard fire.

There are many criteria that allow us to define this resistance, in such a way that they can keep their characteristics in the case of a true fire. For example, we relate the Italian rules about civil buildings in steel.[32]

In Italy the safety regulation for fire protection of buildings with steel structure for civil use classifies the rooms with reference to fire loading.

The fire loading represents the heating potential of the materials which are in the area. It is defined as the total heat produced by the fuel present:

$$Q = \sum_{i=1}^{n} K_i P_i \text{ (MJ)} \tag{1}$$

where K_i is the weight in kilograms and P_i is the heating power in MJ/kg of the ith element of the n fuels present.

The regulation is not referred to the fire loading as defined above but to the specific fire loading, q, measured in wood equivalent kilograms per unit area of the plane surface of the room. To be more specific, the quantity, Q, defined in Equation 1 is divided for the plane surface of the room, Ap, measured in square meters and after it is divided for 18.499 MJ (the heating power of the standard wood), so

$$q = \sum_{i=1}^{n} (K_i P_i)/(18.499 \times Ap) \text{ kg/m}^2$$

The room class is expressed by a number that is the product of the specific fire loading and of a coefficient K, less than one, which takes into account the real condition of risk.

We can find this coefficient in a table and in a diagram in the Italian standard.[32]

This criterion is very practical and useful in the project phase because it brings the numerial identity of the fire resistance to give to the structures with the "room class", and at the same time it is effective for its direct correlation with the fire charge. For example, if the specific fire loading of a room is 60 kg/m² and if we find K = 1 the fire resistance rating must be 60 min or more.

Numerous factors influence the development of the fire and it is important to take each of them into account in the analysis of the various problems of fire defense. However, in many cases the fire loading assumes the main importance because it is the element which affects the duration of the fire more than the others.

2. Fire Reaction

The behavior of materials in the presence of fire is considered in the fire rules with two parameters: the fire resistance already defined and the fire reaction.

The fire reaction represents the degree of participation of the material to the fire to which it is submitted.

At present the fire rules are inclined to assign a class to materials to mean the effective participation to the fire valued by experimental tests.

The characteristic parameters of the fire reaction are inflammability, flame spread rating, dripping, heat development, production of smoke, and harmful substances. The tests try to establish the behavior of the materials in relation to these parameters either at the start or during the fire when the material is highly radiated.[33] Therefore, it is possible to establish the behavior of the material at the beginning of the fire in the case of sparks and how it can contribute to the propagation of a fire which has developed outside the material itself involving this material.

For the building materials it is necessary to know the flame spread over its surface (NFPA 255, ASTM E-84).

The structural components and the thermic and acoustic isolating materials should be noncombustible (see ASTM E-136). The internal finishings and fittings should be certified for an appropriate flame spread rating over its surface (NFPA 255-ASTM E-84).

3. Safety Distances

When it is not possible to limit the consequences of the fire with compartmenting, it is necessary to isolate the area with suitable safety distances.

Three kinds of safety distances are defined: external, internal, and of protection.

The external safety distance is the minimum distance between the perimeter of the fire area and the building or another public structure (a street, a railway, etc.) which is external to the plant enclosure.

The internal safety distance is the minimum distance between the perimeter of the fire area inside the plant.

The protection distance is the minimum distance between the fire area and the plant boundary.

In the first section of this chapter we have written about the problem of thermal effects in relation to the distance from the fire, in order to define the fire areas related to the risk of thermal radiation.

Fire rules state the minimum safety distances; they are defined in such a way as to prevent the fire spreading, but they are unreliable for other effects, such as explosions of ignitible mixture and propagation of explosive clouds for which it is necessary to act with preventive systems within the area.

The safety distances depend on the kind and on the physical state of the fuel stored or used in the fire area. The flammable liquids are subdivided into classes according to the flash point, defined as the minimum temperature at which a liquid gives off vapor in sufficient concentration to form an ignitible mixture. The risk of fire spreading from one building to another depends on liquid class, protection devices and type of tank (see NFPA 30).

B. Active Protection

1. Automatic Fire Detection System

The purpose of automatic fire detectors is to detect in the shortest possible time the presence of a fire in such a way that all necessary measures to fight the fire and to protect the occupants can be taken.

Above all they are used as alarm systems when the fire risk is determined by a high population density or when the goods contained in the area have a great economical value that cannot be compensated; they can be employed as alternative protection with passive defense systems too.

In industrial plants they are employed as alarm systems in offices, in data elaboration centers, and anywhere with the risk of a very rapid development of fire which has to be compensated by a timely warning to obtain an appropriate available time for evacuation of people.

In addition to giving the alarm, automatic fire detectors in industrial activities are employed in conjunction with extinguishing or other protection systems (smoke venting, fire doors, etc.) with immediate or delayed automatic triggering, according to results of risk analysis of the area in question. Depending on cases and purposes that have to be reached, the configuration of automatic fire detection systems can be different. In the flow chart of Figure 8 (from European Standard EN 54 Part 1) the most complete system is reported.

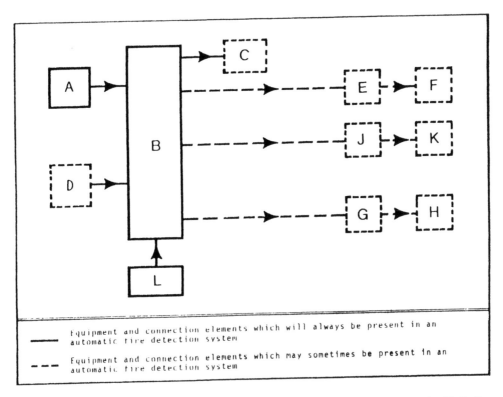

FIGURE 8. Flow chart of an automatic fire detection system: (A) Fire detector, (B) detector control and indicating equipment, (C) fire alarm device, (D) manual call point, (E) fire alarm routing equipment, (F) fire alarm receiving station, (G) control for automatic fire protection equipment, (H) automatic fire protection equipment, (J) fault alarm routing equipment, (K) fault alarm receiving station, and (L) power supply.

With the continuous line we show the components which will always be present, with the dashed line the components which may sometimes be present and that assume functions of particular importance in determinated cases.

The essential components are fire detectors, detector control and indicating equipment, and power supply.

The parts not always required are fire alarm devices, manual call point, fire and fault alarm routing equipment, fire and fault alarm receiving station, control for automatic fire protection equipment, and automatic fire protection equipment.

There are different types of fire detectors. They can be defined according to the physical phenomenon which they detect (heat detectors, flame detectors), their configuration (spot type or line type), or the way they respond to the phenomenon detected (static detectors, if they give the alarm when the magnitude of the measured phenomenon exceeds a preselected value, rate of rise detectors if they operate when the rate of change of the measured phenomenon with the time exceeds a prefixed value).

Regarding criteria of choice of fire detectors, their number, the setting, interconnection of components, and all concerning the realization of the system, it is essential to refer to national standards (NFPA 71, 72a, 72b, 72c, 72d, 72e, 74 and others).

The detectors must be chosen according to the different kinds of fire. The choice is also affected by environmental conditions existing within the building. Fire detectors should also be chosen according to the ceiling height. In general, smoke point detectors give the fastest responses because smoke is almost always the first phenomenon of a fire.

The smoke quickly rises towards the ceiling where it stratifies and, if it goes into the detector, even if it is cold, it sets off the alarm even in low concentration.

The smoke detectors commonly used are subdivided into two types: (1) ionization smoke detector and (2) optical smoke detector.

In the first type the smoke reduces the current flowing through an ionization chamber within which a small radioactive source is situated. The second type is based on the fact that smoke particles produce absorption or scattering of light. Generally when we speak about optical detectors we mean scattering smoke detectors.

The optical detectors cannot always replace the ionization detectors.[34] The latter have a good sensibility for the whole spectrum of smoke that can be produced by different combustibles; while the scattering smoke detectors have good sensibility only for light smokes, they are not very sensitive to dark smoke, and they are not sensitive to invisible smoke. Therefore, the kinds of combustible material existing in the building and the characteristics of the smoke produced by their ignition are essential to the choice of fire detectors. Everywhere that the smoke detectors cannot be used because the fire does not produce visible or invisible smoke, or because the smoke is normally present in the room, it is necessary to use detectors sensitive to other phenomena which occur with the fire, as the heat detector or flame detector.[35]

The number of detectors which must be used depends on the floor area of the protected zone and on other factors that affect the smoke spreading positively if they help its normal path, negatively if they obstruct the reaching of the detection points chosen. Standards generally fix the values of the maximum area protected by a single detector according to height and to the slope of the ceiling. This is the base criterion for stating the number of detectors. In addition there are other criteria that hold in due consideration the geometry of the room, spacing of detectors, irregular areas, heating, ventilating and air conditioning, and the presence of obstacles. The detectors are connected to control and indicating equipment whose aim is to accept the fire alarm signal to monitor the correct functioning of the system, to give audible and visible warning and, if required, to operate protection equipment in order to have a certain indication of that part of the protected area which the alarm comes from. The area must be divided into zones. There is a visible signal for each zone on the control equipment which allows the location of the fire to be found quickly. In fact the operator, once he has reached the zone, will be able to find the primed detector through the optical indication placed on the detector itself.

The detectors in the same zone are connected electrically in one or more circuits. When the automatic fire detection system must operate protection equipment in order to prevent the activation of these systems from being caused by a false alarm, the connection of the detectors in the same zone in two circuits is necessary, In fact when the detectors are connected in two electrical circuits when the first detector is activated it gives only an alarm signal; to set the protection system into action it is necessary that the first detector of the second circuit is operating.

The installation must have at least two completely independent power supply systems, each able to assure the performance of the automatic fire detection system. The primary power supply will be derived from the main electrical system through its own circuit independent from other uses, clearly identified and protected from fire.

2. Control and Extinguishing of Fires by Manual and Automatic System

The main protection is to arrange adequate fire extinguishing systems.

We must distinguish the means intended as a first line of defense to fight fires of limited size (buckets, sand, portable fire extinguishers) from the installation of systems for fire extinguishing (sprinkler, water spray, foam, halon, CO_2, standpipe and hose systems, hydrants, and other fixed protection equipment).

When the quantity and the nature of combustible materials present in the area allow us to foresee a not highly intense fire, but one that could have remarkable consequences for the functionality of systems, quick-action extinguishing substances have to be employed.

Generally these substances will be employed to cope with fire with approved types of extinguishers for localized defense that can be realized, for instance, by automatic extinguishers controlled by suitable detectors.

In any case, portable fire extinguishers containing appropriate extinguishing agents should be located in all the rooms of the industrial plant, even in those provided with a fixed automatic extinguishing system, in order to arrest the fire at its beginning, if it is possible.

If there are electrical tension parts halon or CO_2 extinguishers will be preferred. Standards define classification, ratings, and performance of fire extinguishers (NFPA 10). On the contrary, when the quantity and the nature of combustible materials in the area are such as to presume a large fire it would be better to orientate the choice towards the use of water, when it is possible.

Automatic fire extinguishing systems must use halon or CO_2 in particular cases only, when there is material whose economic value is very high or when the use of water is not allowed because of technical reasons.

Automatic systems using water are sprinkler systems and water spray systems. The first is a system of piping connected to a water supply which, activated by the heat produced by the fire, discharges water over the fire area. The second system is equipped with water spray nozzles which discharge water; it is activated by fire detection equipment. The water spray system must be preferred when there are inflammable liquids and metallic parts that can overheat because of fire in the area. When water is used one must be careful with the electrical voltage parts existent inside the area, in order to avoid electrical breakdowns in the system.

Indoor systems must be predisposed to automatic operation and must transmit an alarm to the control room after they start to operate. They must be dimensioned through hydraulic calculation so as to assure a fixed density in the most unfavorable hydraulic position in the operative area; the density of discharge is determined on the basis of the risk analysis of the fire area. Through the operative area and the density, considering the types of heads and their spacing, we can determine the water supply flow rate and pressure for all systems which must operate simultaneously. The rates at the required pressure should have the capacity to adduce the water to the system for the presumable duration of the fire. The water supplies (water works, pumps, gravity tanks, pressure tanks) must be very reliable. We can also consider very reliable the multiple water supplies on the condition that they are singularly able to deliver necessary flow rates at required pressure.

In addition to the automatic extinguishing systems, external and internal hydrants should be located in all the buildings. The internal hydrants of the wall type, contained in a metallic box, must be equipped with hose of normalized length and with an appropriate nozzle which must allow only a water jet and must be suitable and approved for use on live equipment.

The hydrants must be placed possibly outside the area, near the access doors.[36]

Concerning the rate flow and the autonomy of the water system, the requirements of large storage and of large industries must be held in due consideration. It must be remembered that in some fires in industrial plants the water consumption can exceed 20 to 30 cc/min for over 2 d.

The water supplies (flow rates at pressure) must be planned so as to assure the simultaneous operating of all the nozzles which are considered and which are placed in the most unfavorable position from the hydraulic point of view. The system should be fed by three pumps: (1) an electric motor-driven pump, (2) a diesel motor-driven pump, and (3) a pump driven by such an electric motor that the water systems are at a uniform pressure (compensation pump). Therefore, the system will be dimensioned in such a way as to cope with the fire even if the automatic extinguishing system is out of order; the water supplies (tree pumps system, i.e., the automatic fixed extinguishing systems) must have their own independent supply.

The main pipeline must have a ring-like shape provided with an interception value so that it allows maintenance without interrupting the supply to the hydrants.

In some industrial activities, with presence of inflammable liquids it is necessary to dispose of manual or installed means for the production of foam and, in some cases, to organize foam extinguishing systems (NFPA 11-NFPA 11A-NFPA 11C).

In such cases it is necessary to design the size of systems and quantity of liquid foam according to the characteristics of the system in such a way as to have, in a short time, the necessary depth of foam over the whole surface to be protected.

The areas protected by hydraulic systems, either manual or installed ones, must be pre-disposed in such a way as to reduce to a minimum the breakdown caused by the water discharged by the systems themselves towards the building and its content. To this purpose all the means able to avoid dispersion of water, to gather it, and to convey it to a suitable place as soon as possible must be set up.

The available flow rate should be at least equal to the flow rate of the total discharge of the plant on the operative area, or on the maximum collection basin that can be verified inside of it, adding a quantity equal to the one delivered by the hydraulic nozzles foreseeably operative on the area.

If there are only hydrants the calculation of the flow rate must be effected considering nozzles which are operating simultaneously.

3. Smoke-Venting Systems

The production of smoke makes the consequences of fires much worse. In fact smoke is the most frequent cause of death, because of the toxic materials that it contains. It is a vehicle of fire propagation due to the velocity with which it spreads in the vertical and horizontal directions at a high temperature, and to the incandescent particles it carries along its path; it leads escaping people into panic and makes the rescue operations and the extinguishing difficult.

Against these obnoxious effects large flows of ventilation that aim to rid of smoke the lower parts of the rooms in which the fire has developed and to avoid the horizontal propagation by vertical venting and direct expulsion towards the outside must be realized.[38]

The smoke-venting systems find their most useful and effective application in large un-divided floor areas with high ceilings, which are frequently found in industrial buildings.

In this case the system is constituted by:

1. Vents that are openings on the roof which, during fire, allow direct communication with the outside, or openable structures on the face of the building that are openings similar to the preceding ones but placed on the vertical external walls, in the high parts of the room, or finally small openings for extraction that expel smoke towards the outside by means of pipelines.

2. An air inlet system that can be constituted by the admittance doors themselves or by openable structures on the face situated in the lower part of room or by suitable small inlets.

3. Curtain boards (that are noncombustible) and fire-resistant vertical panels (that descend from the ceiling) subdivide the area into sectors inside which the smoke must be confined, preventing in this way its propagation in a horizontal direction. The curtain boards cannot be adopted when the area is not large enough or does not extend too much in a longitudinal direction. The correct design of the system determines a zone free of smoke H^1 above which the volume invaded by smoke develops; this volume does not always assume a well-defined geometrical shape. For its evaluation one should refer to a covering plane H-m high, that is, as high as the arithmetic average between the height of the highest point and the height of the lowest point of the ceiling (see Figure 9). When the curtains are employed the rooms are subdivided into sectors whose maximum surface in plan is fixed by standards. The smoke-free height is fixed by the

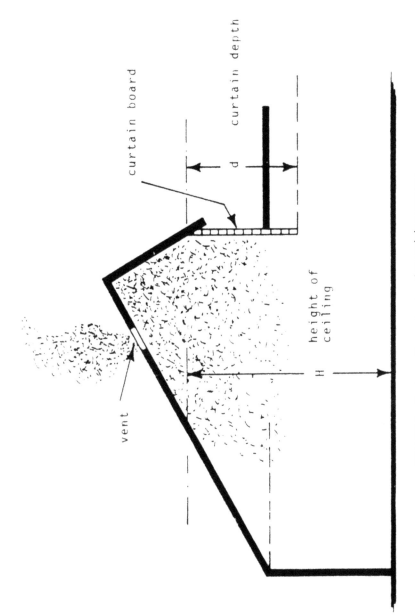

FIGURE 9.   Smoke venting systems — characteristic parameters.

development of the curtain towards the bottom and generally it does not go down 2.50 m under the ground. If there are no curtains the sector is constituted by the entire room.

The vent area is a part of the sector surface in plan, a sector that can be obtained from standards in function of the average height H, fire growth time, curtain spacing, and minimum clear visibility design time (NFPA 204 M).

Concerning the letting in of air, it must take place directly from the outside or from sectors communicating with the outside and the inlet surface must never be smaller than that for extraction.

The vents must be located so as to allow the smoke flux to move freely towards the outside without obstacles and without causing danger of propagation of fire.

In several-storied buildings natural ventilation can still be applied by arranging small openings for letting in and for extraction for each floor, linked to one or several vertical pipelines directly communicating with the outside.

Criteria of setting up and designing are similar to those fixed for a large floor area. Usually, however, in several-storied buildings, the mechanical ventilation is more practical. In this case the smoke extraction takes place by means of fans located on the top of the buildings, equipped with a resistance to high temperatures, at least 400°/60 min, and built-in noncombustible material also in the accessories which are linked to the pipelines.

The letting in of air can be natural from the outside or from the perimetrical sectors, as in the case of natural ventilation or by means of fans which let air at a limited speed into the various rooms through small openings.

Either the small extraction openings or the inlets are equipped with air locks required not only in one-floor buildings.

Even though the ventilation is forced the subdivision into sectors by curtains can be necessary.

When mechanical vents are adopted the total vent area is replaced by conversion from vent areas depending on the depth of the curtain board.

We must give particular importance to the protection from smoke of the staircases. On staircases smoke extraction by ventilators is never allowed; therefore, a natural ventilation system with 1-m² large vents located on the roof with inlets of the same area at the bottom is adopted, the clean air of the staircases can go into the other rooms, but the air of the floor involved in the fire can never invade the staircase through the entrances. The control devices must open the vents and air inlets at the right moment and at the same time they must cause the ventilation to stop when this has been built separately from the smoke-venting system.

The control devices must be located in easily accessible places for intervention and which can be closed off from protected rooms when they are no longer necessary. The activation can be manual, pneumatic, electric, hydraulic, or mechanical. The activation can also be automatic, preferably operated by an automatic system of smoke detectors, but also in this case manual activation must be possible.

V. MITIGATION OF EFFECTS OUTSIDE THE FIRE AREAS

A. Boundary Conditions

The fire areas, as defined in the beginning of this chapter, must assure that a fire which has broken out inside them should not produce effects outside and vice versa every fire that breaks out outside these areas must not have consequences inside of them.

Some boundary conditions must be verified realizing the following points:

1. If the fire area expands outside the buildings there must be a safety distance between the area and the buildings so as to avoid the spreading of the fire and its effects outside the area. These distances can be valued by the calculation methods previously proposed.
2. The fire areas inside the buildings must satisfy different conditions. The fire load must be consistent with the fire resistance of the structures defining the area, in order to prevent the fire spreading outside. A similar check must be carried out in order to prevent the fire from spreading on the inside. It is a good rule to assume 180 min as the value of the fire-resistance rating. The fire resistance of structures of the fire areas should be determined, in the design phase, by calculation methods based on the distribution of temperatures inside them, with the hypothesis that they are exposed to the time-temperature standard curve and verify the stability conditions in correspondence to the variations of the stresses and of the constraint conditions. The calculation of the temperature distribution inside the structures as function of time under the hypothesis of a standard exposure to the fire allows us to determine the variations of temperature and of the physical characteristics of materials constituting the structures (yield point, modulus of elasticity etc.) in order to value the limit of breakdown of the supporting structures and the reaching of 150°C in the unexposed bands of the separating surfaces.
3. The penetrations of the plants must not reduce the fire resistance of the wall and the slab on which they are made. The verification of this condition must be carried out by the time-temperature standard curve. In addition the penetrations made on structures delimitating rooms at different pressures must be homologated by tests reproducing the real conditions. In the pentrations the sealing should be executed in noncombustible materials.
4. The doors made on fire-resisting walls and eventual air-lock crossing walls or slabs of fire areas can be disregarded when considering the continuity of the structure, on the condition that they are operated by an automatic and manual system that guarantees their closing in case of fire and that if submitted to the fire endurance test, effected according to the time-temperature curve, they present a fire-resistance rating equal to that required for the structure which they are made on. The windows on the structures which define the outline of the fire areas should not allow the fire, eventually produced in the area, to spread to another fire area. The materials of the structures and of the constructive elements defining the fire areas should be noncombustible.
5. Drainage planned for draining off of the extinction water should be realized in order to avoid the spreading of the fire caused by the presence of inflammable liquids.
6. Eventual short circuits or damage to cables by the fire must not provoke malfunctioning, wrong maneuver, or false indications outside the fire area where the cables are located.

VI. SAFETY OF THE OCCUPANTS

A. Classification of the Occupants

The occupants involved in the fire may be distinguished as follows:

1. The people that must be evacuated
2. The technicians whose presence is requested in case of emergency for the safety of the plants
3. The staff intervening against the fire and rescue teams

People who occupy the fire areas must be evacuated and must reach a safe place within a time short enough to prevent them from suffering any damage.

Therefore, a system of emergency exits must be planned so as to allow the occupants to go out quickly and in an orderly way. This problem will be dealt with in the following sections.

People that are necessary to direct the emergency must be able to work in safe conditions.

Therefore, the presence of smoke and of toxicant, radioactive, and inflammable materials in the areas where such persons must intervene in the case of fire must be avoided by the installation of a ventilation plant. In particular the control room must be planned and located so as to offer the personnel the best protection against every harmful event (overpressure, heat radiation, and nuclear radiation) and to allow the occupants to escape in case of necessity.

If necessary, the ventilation plant could be made up of multiple systems, of pairs of blowing and sucking ventilators which, provided with air locks, allow us to control the propagation of the fire and the release of dangerous substances. In the rooms where, due to the nature and to the quantity of fuel, a large quantity of smoke could develop in a case of fire, smoke and heat venting must be used.

Smoke-venting systems must be made also in order to help the staff in fire fighting. In the presence of smoke the fire department should be able to enter the area and fight any fires in central portions of the building. Entrances for succorring vehicles must be placed along the building perimeters so that they can enter places near the fire areas.

Also suitable passageways must be made in order to put the fires out manually.

1. Evacuation

In order to define suitable systems that make easy the evacuation of the occupants in the case of fire, the NFPA 101 standard can be followed. We distinguish two cases: (1) the areas where the people are few and (2) the crowded areas.

2. Low Density of Employed Population Area

The time available for evacuation starts as soon as people realize they are in danger and finishes when the development of the fire has produced smoke, toxic substances, and heat in such quantities that there is no longer any chance of life in the area.

It is necessary that the effective evacuation time, i.e., the ratio between the distance covered during the escape and the velocity of escape, is shorter than available evacuation time.

If there are few people in the room and the velocity of escape does not depend on their number, such a condition can be satisfied to make the way out easy and shorten the distance to the outside. The means of exit should be protected from the smoke and the heat developed from the fire; they should consist of principal walkways which avoid any blind passages and tunnels. The floors and the staircases must be antislip, the building materials must be noncombustible, and they must not develop smoke and toxic products under heat action. The means of exit must be provided with independent illumination systems and be protected from the fire and there must be clear indications of the emergency exits.

The emergency exits must be large enough, conveniently located according to the way of exit, and provided with suitable doors. Sharp corners and other conditions which can obstruct the evacuation should be avoided. For shortening the distance to the outside, some safe places should be realized.

Safe places are open places or fire compartments that are spaces within the building enclosed by fire barriers. They must be suitable to receive and contain a predetermined number of people or to allow them to move out ordinately.

The fire areas, when they are fire compartments, can be considered intrinsically safe places so that the means of escape towards other fire areas can be allowed. In this case the emergency exit must lead into the safe place through a "smokeproof enclosure". The smokeproof enclosures are rooms designed with structures and doors that are fireproof. The doors are provided with self-closing devices (Figure 10).

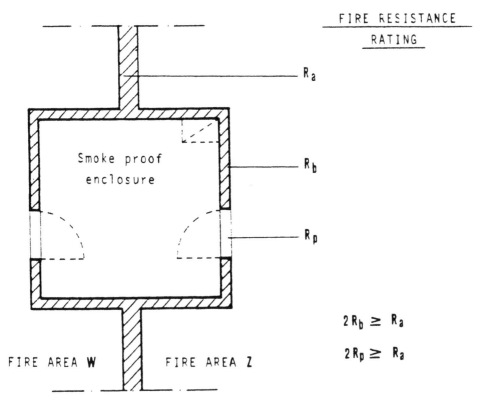

FIGURE 10.   Smokeproof enclosure.

Protection from the smoke is guaranteed by ventilation stock aeration from the outside through openings or also by keeping the rooms overpressured in comparison with the adjoining rooms also in a condition of emergency.

3. High Density of Employed Population Area

In the areas where many people are foreseen to be present, the speed of escape is influenced by the people themselves.

In this case the evacuation time is not proportional to the distance and defining the capacity of means of exit one must consider a parameter, the capacity per unit of width, which is proportional to the evacuation time, but which considers also the decrease of speed of escape caused by the presence of people.

The capacity per unit of width, C, is defined as the largest number of people that, in a means of exit, are supposed to be able to go out through the unit exit width (the unit exit width is 22 in. according to the NFPA Standard).

The capacity per unit of width depends on the evacuation time T_e as follows:

$$C = c \times T_e$$

where c is the specific capacity per unit of width, i.e., the largest number of people who are able to go out through the unit of width in the unit of time.

The specific capacity, c, depends on the linear density D_l defined as the number of people present in 1 m and decreases if the linear density increases.

Data concerning the circulation on a level, in normal conditions, directly determined by experimental observations, are reported in Table 1. You can see that the product of the escape velocity times the linear density is the specific capacity c:

Table 1

Escape velocity (m/s)	0.9	0.65	0.5	0.4	0.3	0.25
Linear density (pers/m)	1.5	2	2.5	3	3.5	4
$C = D_l \cdot V$ (pers/s)	1.35	1.3	1.25	1.2	1.05	1

$$D_l \times V = c \text{ pers/s}$$

This product expresses the relation between the travel distance L and the capacity per unit of exit width, the two main parameters the planner disposes of to decide upon the system of the exits, which allows the orderly flow of people within the fixed time T_e.

If, for example, one considers a capacity per unit of exit width C = 60 and the distance L = 30 m, one obtains:

$$D_l = C/L = 2 \text{ pers/m}$$

because of that, in horizontal exits, one can suppose V = 0.65; it follows that T_e = 30/0.65 = 46 s.

In the various cases truer values of the capacity per unit of exit width and of travel distance can be chosen.

In very large areas it can be convenient to choose higher values of the travel distances and of the capacity per unit of exit width.

If, in the preceding example, we suppose L = 40, we will have D_l = 60/40 = 1.5 then v = 0.9 and T_e = 40/0.9 = 45 s.

The highest values of the travel distance and of the capacity per unit of exit width are fixed by the fire rules.

On the basis of the criteria just explained, the planner will be able to find the values in unforeseen cases.

Once the capacity per unit of exit width is fixed, the total width of the emergency exits is known; in fact the total width of the emergency exits is determined by the ratio between the largest allowed crowding on that floor and the capacity per unit of the exit width. The total width is then divided among several doors, conveniently located.

The largest permitted crowding on one floor can be determined by dividing the floor surface in square meters by the occupant load in square meters per person allowed by fire rules for the various occupancies.

4. Staircases

When the means of exit comprise staircases, different solutions connected with the building height are provided.

In very high buildings, smokeproof staircases are required.

A smokeproof staircase is inside a fire compartment whose entrances, one for each floor, are provided with smokeproof enclosures (Figures 11 and 12).

If the buildings are not very high (not higher than three floors), simply protected staircases are sufficient. A protected staircase must be inside a fire compartment and there must be an entrance on each floor, but not through a smokeproof enclosure (Figure 13).

Ordinary staircases are allowed only up to two stories above ground; in this case one must consider the length of the ramps when calculating the travel distance to the outside.

The staircases must allow a quick evacuation and so they must have particular characteristics. The staircases and the landings must be as large as the emergency exit of which they are a part. If the staircases pass through several floors, to avoid crowding on the lower floors and not to obstruct the escape velocity, the width of the staircases must be chosen considering the capacity of unit of exit width of the several floors.

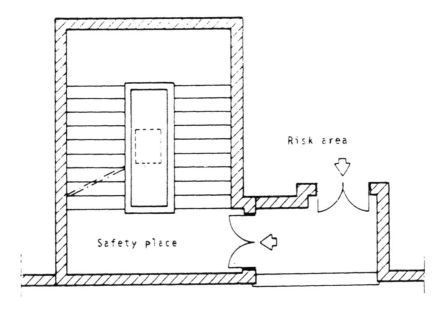

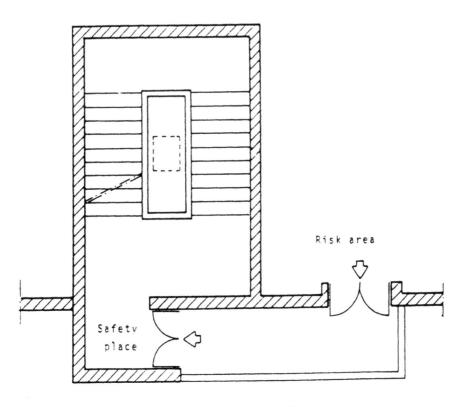

FIGURE 11. External smokeproof staircases.

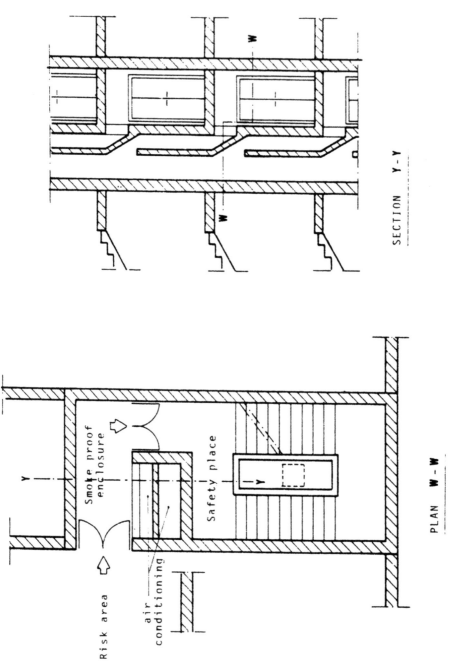

FIGURE 12.   Internal smokeproof staircases.

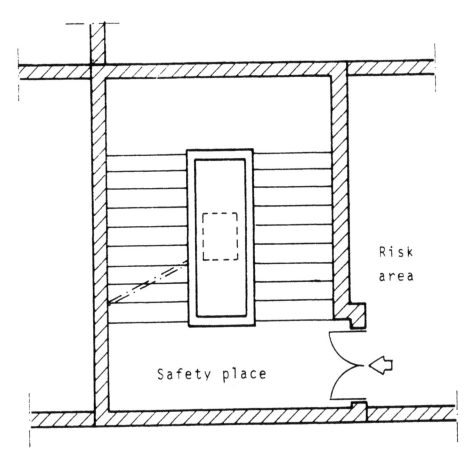

FIGURE 13.   Protected staircases.

REFERENCES

1. **Hottel, H. C.,** Certain laws governing diffusive burning of liquid, Fire res., *Abstr. Rev.,* 1, 41, 1959.
2. **Burgess, D. S. et al.,** Diffusive burning of liquid fuels in open trays, Fire res., *Abstr. Rev.,* 3, 1977, 1961.
3. **Thomas, P. H.,** The size of flames from natural fires, 9th Int. Combustion Symposium, 1963.
4. American Gas Association, LNG Safety Research Program, Rep. 15-3-1, 1974.
5. **Moorehouse, J.,** Scaling criteria for pool fires derived from large scale experiments. I, Chem. E. Symp. Ser. No. 71, 1982.
6. **Mudan, K. S.,** Thermal Radiation Hazards from Hydrocarbon Pool Fires, *Prog. Energy Combust. Sci.,* 1984.
7. **Hottel, H. C. et al.,** *Radioactive Transfer,* McGraw-Hill, New York, 1967.
8. **Hawtorne, W. R. et al.,** Mixing and combustion in turbolent fuel jets, 3rd Symp. Int. Combustion, Baltimore, 1949.
9. **Sunnavala, P. D.,** Dynamics of the buoyant diffusion flame, *J. Inst. Fuel,* 40, 482, 1967.
10. **Sunnavala, P. D.,** Determination of flame length for freely burning laminar and turbolent flames, *Chem. Age India,* 11, 217, 1960.
11. **Baron, T.,** Reactions in turbolent free jets — the turbolent diffusion flame, *Chem. Eng. Progr.,* 60 73, 1954.

12. **Raj, P. K. and Kalekar, A. S.,** Assessment models in support of the hazard assessment handbook, U. S. Coast Guard Rep. CG-D-65-74, 1974.

13. **Lees, F. P.,** *Loss Prevention in the Process Industries,* Butterworths, London, 1980.

14. **Faj, Y. A. and Lewis, D. H.,** Unsteady burning of unconfined fuel vapor clouds 16th Inst. Symp. Combustion, Pittsburgh, 1976.

15. **Hasegawa, K. and Stato, K.,** Experimental Investigation of the Unconfined Vapor Cloud Explosion of Hydrocarbons, Tech. Mess. Fire Research Inst., No. 12, Japan, 1978.

16. **Mudan, K. S. and Desgroseillers, G. J.,** *Thermal Radiation Hazards from Hydrocarbon Fireball,* ADL, London, 1981.

17. **Markstein, G. H.,** Radioactive energy transfer from turbolent diffusion flames. *Combust. Flame,* 27, 51, 1976.

18. **Bello, G. C. and Romano, A.,** *Cetra, un Codice di Simulazione dell'Irragiamento Termico da Fireball,* Tema, Milano, 1982.

19. **Brzustowski, T. A.** Flaring in the energy industry, in *Prog. Energy Combustion Science,* Vol. 2, Pergamon Press, Elmsford, NY, 1976, 129.

20. **Cagnetti, P.,** Contaminazioni ambientali e dosi alla popolazione — Atti del Convegno "Interventi Antincendio in presenza di sostanze radioattive", Rivista Antincendio No. 12, 1983.

21. **Bello, G. C., Romano, A., and Galatola, E.,** *Soprano, un Modello de Simulazione del Rilascio e Dispersione di Gas da Recipiente a Pressione,* Tema Milano, 1985.

22. **Fryer, L. S. and Kaiser, G. D.,** Denz, a computer programme for the calculation of the dispersion of heavy toxic gases in the atmosphere, SRD R152, UKAEA, 1978.

23. Fire protection guidelines for nuclear power plants, USAEC Regulatory Guide 1.120, U.S. Nuclear Regulatory Commission, Washington, D.C., 1977.

24. **Rasbash, D. J. et al.,** Design of an Explosion Relief System for a building handling liquified fuel gas. Process Industry Hazards, Int. Chem. Energy Symposium Ser. No. 47.

25. **Lunn, G. A.,** Venting gas and dust explosions, A. Review, IChemE, 1984.

26. **Runes, E.,** Explosion venting - loss prevention, Proc. 6th Symp. Loss Prevention in Chemical Industry, New York, 1972.

27. **Mazzini, F. and Pelagatti, M.,** Impianti elettrici nei luoghi con pericolo di esplosioni o di incendio, EPC, Rome, 1986.

28. Comitato Elettrotecnico Italiano, Norme CEI, 64-2, Milano.

29. **Magnusson, S. E.,** Temperature-time curves of complete process of fire development, in Acta Polytechimica Scandinavica, C1 65, Stockholm, 1970.

30. **Cuomo, S.,** Resistenza al fuoco delle strutture e sua determinazione, EPC, Rome, 1968.

31. Relazione finale della Commission di studio per le norme per la protezione contro il fuoco nelle costruzioni a struttura in acciaio, Bollettino Ufficiale CNR (norme tecniche) A VII No. 37, 1973.

32. Ministero dell'Interno, Norme di sicurezza per la protezione contro il fuoco dei fabbricati a strutture in acciaio destinati ad uso civile, Cicolare No. 91, Italy, 1961.

33. Ministero dell'Interno, Classificazione di reazione al fuoco ed omolagazione dei materiali ai fini della Prevenzione Incendi, Italy, 1984.

34. **Mazzini, F.,** Impianti di rivelazione automatica di incendio con impiego di rivelatori di fumo puntiformi, Rivista Antincendio No. 7, Rome, 1983.

35. **Pasquinelli, M.,** Rivelatori di calore, Rivista Antincendio No. 7, Rome, 1983.

36. **Amore, P.,** Criteri di progettazione di una rete idrica antincendi con idranti. Rivista Antincendi con idranti. Rivista Antincendio No. 3, Rome, 1960.

37. Concordato Italiano Incendi, Sistemi per l'evacuazione dei fumi "Criteri di costruzione ed installazione", Milano, 1977.

38. Home Ministry, About the smoke evacuation in the very high buildings, Standard 7 giugno 1974.

Chapter 4

ARTIFICIAL EVENTS: CLOUDS EXPLOSIONS, DANGEROUS CLOUDS, AND PHYSICAL PROTECTION

Salvatore Ragusa

TABLE OF CONTENTS

I. VAPOR CLOUD EXPLOSIONS

A. Cloud Formation and Ignition

Every time large quantities of gaseous combustibles are released, there is high chance of having an unconfined vapor cloud explosion (UVCE). The release, for instance, can follow a runaway reaction resulting in the bursting of the reaction vessel, or come from a leakage, or from an explosion of a liquid container where it has been filled without leaving any vapor space; cubic thermal expansion coefficients in liquids are 20 to 30 times greater than in steel and so heatings of only a few degrees (°C) are enough to exceed the steel-breaking stress of a sealed fully filled container.

In fact, after a containment barrier rupture the combustible released can be kept in a confined space or spread in free atmosphere, depending on circumstances. In the first case, a rich mixture is formed and generally it will end up by burning. The flame propagation will be laminar, usually without explosion. In the second case a veritable vapor cloud is formed and there are three possibilities: (1) the dispersion goes on until the mixture becomes so poor to not offer any risk, (2) the mixture is ignited and burns slowly, and (3) the ignited mixture burns so rapidly to generate explosive waves. The waves propagate in the vapor cloud and, arrived at its boundary, move forward in the atmosphere (the support energy is produced in the thin layer of the combustible zone).

Normally, the ignition source is founded in a short time and not far from the release area. However, delays of 15 min and distances of several hundreds of meters or of kilometers cannot be excluded.[1-3] In 81 accidents concerning tank trucks, 58% of the releases found an ignition source nearer than 15 m, 76% nearer than 30 m.[1] In 150 accidents 69% of them had the ignition before 1 min from the release, 83% before 5 min, and 95% before 30 min.[4]

There are no well-defined theories to foresee if the explosion of a combustible unconfined mixture will be a deflagration (flame velocity less than sound velocity) or if it will be a detonation (flame velocity more than sound velocity) whose effects are much more serious. Detonation can be reached in two ways. In the first one the flame, starting at low velocities, accelerates because of the turbulence created from itself or because of existing obstacles in the still unburnt mixture. If the flame has enough space to travel (large clouds) it continues the accelerate motion and the sound velocity can be attained and surpassed. In the other way the detonation is directly reached without having a deflagration before. This is possible only when the ignition source is very strong; the source must trigger a strong shock wave and then support it for a sufficient duration.

Normally, the provisions about final flame velocity, i.e., the provisions if a detonation can be reached or not, are connected to the dimensions of the cloud. According to statistical observation, in some safety analyses the following criterion has been adopted:[5] if the flammable substance contained in the cloud is of the order of 100 tons, a probability of 0.1 is attributed to a detonation (or to a fast deflagration). For flammable gases clouds of the order of 1000 tons, the detonation is considered as a sure event.

B. TNT Equivalence

The development of an explosion is conditioned from the reaction piloting the shock wave. Limit situation is that caused by a so-called "ideal" explosive, that is, a very high energy density punctiform material which instantaneously releases the whole available energy. Conventional explosives, as the trinitrotoluene (TNT), are not so far from ideal ones.

As indicated in Figure 1, in an ideal explosion pressure wave at time t_1 reaches the point P in the surrounding space. There, the pressure from the initial value p_0 raises instantaneously to p_1. Then the pressure in P falls to p_2 and finally it comes back to the primitive value p_0. Therefore, two phases can be distinguished: the first one with positive pressure wave and the other one with negative pressure. While the negative phase is nonimportant in explosives,

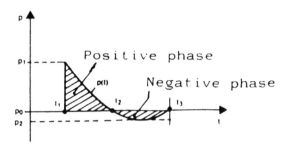

FIGURE 1. Ideal source: explosion pressure vs. time.

Table 1
PROPORTIONAL LAWS FOR IDEAL EXPLOSIVES

Explosive mass	Time	Distance	Pressure peak	Positive phase length	Impulse
m	t_1	x	p_1	$t_2 - t_1$	i
$\alpha^3 m$	αt_1	αx	p_1	$\alpha(t_2 - t_1)$	αi

in other cases it can be remarkable. The upper area between the p_0 line and the pressure curve is the positive impulse i:

$$i = \int_{t_1}^{t_2} [p(t) - p_0]dt$$

Within certain limits, the characteristic quantities of ideal explosives are regulated by the proportional laws of Table 1. It means that if m kilograms of a given explosive produces at a distance x of 100 m a pressure peak p_1 at the point P, 2m will produce the same pressure peak at a distance of

$$100 \cdot 2^{0.333} = 126 \text{ m}$$

UVCEs show a quite different behavior from an ideal explosion, essentially because when a cloud explodes small energies per volume unit are released in relatively important times. But an equivalence with TNT is accepted in the far field. In the intermediate and in the near field the equivalence is not acceptable, particularly when the cloud is too lengthened; in this case an equivalence with the TNT would give an important overestimate of the pressure peak in the near field and an important underestimate in the intermediate field.[6]

In an equivalence assumption, the same damages are provoked from the mass m_s of the combustible substance present in the cloud and from the mass m_{TNT} given by the expression:

$$m_{TNT} = \frac{\eta m_s E_s}{4.25}$$

Here, E_s [MJ/kg] is the energy developed in the complete combustion in air of 1 kg of the substance; 4.25 MJ is the energy developed in the combustion of 1 kg of TNT; η is the equivalence factor (or the explosion yield) which takes into account the cloud formation and diffusion processes and the real expansion work related to the whole energy content in the cloud.

On the basis of the effects of 21 accidents the relationship of Table 2 between explosion

Table 2
UVCE EVENT PROBABILITY VS.
EXPLOSION YIELD

Event probability	Explosion yield(%)
0.51	≤ 4
0.76	≤ 10
0.82	≤ 12

yield and event probability of such a yield have been given. In safety analyses η is generally taken equal to 0.1; in this way it can be assumed that, in the far field, 1 kg of hydrocarbons (that normally develops 42.5 MJ) roughly gives the same damages as 1 kg of TNT does.

A recent experimental work concerning optimal spherical air mixture of hydrocarbons (ethylene, acetylene, propane) gave the following expressions for the overpressure peak p_1 and for the impulse i [bar ms]:[7]

$$\log(p_1 - p_0)/p_0 = 0.3527 - 1.8187(\log xE^{-0.333}) +$$

$$0.2410(\log xE^{-0.333})^2 - 0.0320(\log xE^{-0.333})^3$$

They are applicable in the range $0.25 \leq xE^{-0.333} \leq 25$ in which x is the distance (m) and E [MJ] the energy released in the explosion and are applicable in "free air" explosions; near surfaces, or soil, reflected waves locally create a worse situation.

C. Hazards from UVCEs

Methane and liquefied natural gases (LNG) are much less dangerous than liquefied petroleum gases (LPG); the first causes only 10% of fires caused by LPG.[8] Generally speaking, flammable gases with two to six carbonium atoms are responsible for the more serious accident in UVCE.

Explosions can be accompanied by fires. In 170 cases of UVCEs 35% had an explosion as main consequence, 29% both fire and explosion, and 34% fire alone.[9]

The burning mass can assume the form of a fireball. Depending on the quantity of combustible mixture, the diameter of this ball can reach hundreds of meters. Temperatures are greater than 1000°C. Before collapsing, the fireball radiates important amounts of thermal energy for tens of seconds, even at distances of kilometers.[10]

In seven representative accidents happened in Europe and in U.S., released combustibles being of the order of some tens of tons, more than 300 people were killed: an average of 1.2 people for each ton released.[11]

The radius x separating fatal and nonfatal injury area around the center of an explosion has been conservatively suggested as

$$x = 5m^{0.333}$$

where m (kg) is the mass of the explosive substance released.[12] Within the dangerous area around potential risk source any concentration of people should be avoided. Office blocks, dining centers, and so on should be built out of these boundaries.

In case an UVCE occurs, man can be affected from:

- Pressure variation,[13,14] critical organs are liver, lungs,[15,16] and ear drums[16,17]
- Wounds and trauma from missiles[13-15,18]
- Knocks of human bodies on walls, obstacles[13,14]
- Burns of retina and of skin in the associated fires[19,20]

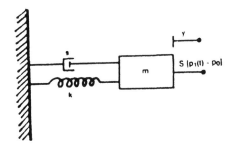

FIGURE 2. Dynamic load on structure, schematization.

D. Shock-Wave Effects on Structures

Even if the overpressure wave form in a UVCE is typical of each explosion, usually TNT blast wave is taken as reference. Generally, negative phase is neglected; hence the load and the impulse on the structure are only determined from the positive phase. This load is

$$Sp(t) = S(p_1 - p_0)e^{-t/T}$$

The positive impulse is

$$I = S(p_1 - p_0) \int_0^\infty e^{-t/T} dt$$

In the preceding formulas, S represents the struck surface and, assuming that the overpressure follows an exponential law, T represents the period (time needed to the overpressure to fall at 1/e of its peak value). Then the charge is a dynamic one, changing along the time; the structure answer, deformation, and stress will change too along the time. The structure will assume a vibrational state but soon it will come back to its initial state and position (if the phenomenon is kept in the elastic range) after dissipation by heat of the deformation energy previously absorbed.

In a very simplified approach, let us consider an elementary structure, a restrained beam, whose mass is concentrated in a point. To verify its resistance the mass displacement is determined and, by the displacement, the stress is founded. Supposing that the beam mass can move in only one direction, a simple motion equation is obtained. Such a configuration is sketched in Figure 2 where all the fundamental system properties are expressed: the mass m, the elastic strength to the displacement, represented by a spring whose stiffness is k, the dynamic load, the displacement y proportional to the elastic force, and the energy loss process, represented by a damper having a coefficient s. Usually, damper forces are supposed to be proportional at the velocity and in opposite direction to motion. When the damping coefficient is smaller than the so-called "critical" damping, as generally happens in real structures, the beam will again assume the initial configuration after a series of damped oscillations.

The complete differential equation giving the force balance is

$$m\ddot{y} + s\dot{y} + ky = S(p_1 - p_0)e^{-t/T}$$

The damping effect, whose contribution is normally only a few percent, can be neglected $(s\dot{y} \simeq 0)$. The solution gives the maximum elongation of the beam. It is possible to represent this elongation as a function of the beam vibration frequency ω, this being equal to $k^{0.5}/m^{0.5}$. Therefore, the maximum elongation y_{max} is

$$\text{for} \quad \omega T \ll 1 \qquad y_{max} = \frac{S(p_1 - p_0)}{k} T$$

$$\text{for} \quad \omega T \gg 1 \qquad y_{max} = \sim 2 \frac{S(p_1 - p_0)}{k}$$

Remembering that the structure elongation in static conditions is $S(p_1 - p_0)/k$, we can realize that in dynamic condition:

1.   If $\omega T \ll 1$, the maximum elongation (and the maximum stress) is only a fraction of the value reached in static conditions because the load is taken off before the structure has had enough time to undergo a meaningful deformation.
2.   If $\omega T \gg 1$, the maximum elongation (and the maximum stress) is twice the value reached in static conditions.
3.   In intermediate situations, when the pressure wave period T has value of the same order of $1/\omega$, the maximum elongation (and the maximum stress) is not far from values reached in static conditions.

In a more realistic approach, it must be considered that the structure hit from an important explosion wave can stand beyond elastic limits. Therefore, its elastic-plastic behavior has to be considered. Then, relationships between elastic force displacement and damping force velocities will not ever follow a proportional law; mathemetical solutions will be more complex; finite elements methods are needed in order to describe the structures with models that are nearer to real configurations.[21] The masses of the different structure components are distributed in discrete numbers; material characteristics are fully considered. However, more approximate solutions are also available and presented in a graphic form.[19]

Tests on model are a complement or a possible alternative to calculations. Blast effects of TNT charges are gathered in "isodamage" curves (e.g., permanent deformation of piping having a given diameter and a given thickness, breaking of 10% of glass or of 90% of internal walls, and so on). Methods to reduce the number of tests are suggested.[22]

If a structure can be involved in an external UVCE, the current trend on its design philosophy can be summarized in this way:[23,24]

1.   Structural damages must not affect the personnel safety.
2.   Control rooms, electrical rooms, buildings containing vital equipment or dangerous substances have to face with success shock-wave consequences.
3.   Safety equipment (instrumentation, logics, safety devices) must maintain its efficiency during and after the accident.
4.   Generally speaking, the structure itself can be defined as resistant to an explosion if it can hold up the effects of a 30-m far mass of 1000 kilograms of TNT located near to soil surface.
5.   In a probabilistic approach, a structure could stand "major" stresses once in its life, as long as after the explosion it still keeps its strength performances for "normal" stresses.

II. DANGEROUS CLOUDS

Conditions created from explosions and/or from fires associated with explosions can be worsened by the presence of particular materials dangerous to the health. Let us consider toxic gases, poison substances, and radioactive materials.

A. Toxic Gases

Due to the combustion, carbon and/or other substances contained in combustible material could develop reaction products, the so-called "fire gases" of such nature and amount to be even lethal. Formation of these gases is affected by many parameters, such as the chemical composition of the burning materials, the temperature, and the amount of oxygen present.

Carbon forms carbon monoxide and carbon dioxide, in a mutual proportion depending on the available oxygen. Carbon monoxide is dangerous even at low concentrations: 10,000 ppm are considered to be fatal when inhaled for 1 min, the allowable exposure for several hours being 100 ppm only.[25] Carbon dioxide is dangerous also; it stimulates the inhalation rate of air (which is mixed with toxicants and irritant substances) making breathing more difficult. The deficiency of oxygen by some percent, caused by the consumption during combustion, can cause trouble too.

The other reaction products can have important toxicological effects, e.g., halogen and gases obtained from combustion of PVC (HCl), of fluorinated resins (HF), of some fire-retardant materials (HBr); hydrogen cyanide (HCN) obtained from combustion of paper, polyurethane, wool, nylon; acrolein ($H_2C = CHCHO$) obtained from incomplete combustion of oils; sulfur dioxide (SO_2) obtained from combustion of materials containing sulfur. Short time expositions (10 min) to concentrations of the order of 100 ppm to these fire gases could be lethal.[26]

B. Poisonous Substances

Poison containers or process components containing poisons can be destroyed or damaged from explosion or fire. As a result, poisons could be sprayed out and constitute a real danger to plant operators. Among the poisons we must include pesticides, able to kill any form of life, humans comprehended. Moreover, solid pesticides are usually dissolved in flammable or combustible liquids in order to make easier their application.

Usually, the acute toxicity is given in terms of LC_{50} and LD_{50}, expressed as milligrams of dose per kilogram of body mass of test animal. The term LC_{50} means the concentration in air of vapors, mists, fumes, and dusts which is lethal to half (or more) of the animals used for the test. Similarly, the term LD_{50} is used for liquid and solids swallowed or absorbed through the skin. Poisons can be classified as extremely toxic, highly toxic, or simply toxic. As an example, materials having LC_{50} of 0.1 mg/l (rat) or LD_{50} of 5 mg/kg (rat, i.e., 0.35 g in a 70-kg man) are considered extremely toxic. (See Chapter 2 for further details).

In Table 3 many poisons and pesticides are listed. Some of them can cause serious illness or death by external bodily contact, or when swallowed or inhaled in small quantities. A few, when taken orally, can be lethal in doses less than 10 drops. (See Chapter 8 for further details)

C. Radioactive Substances

Radioactive substances have identical fire and explosion hazards than the same material when not radioactive. More, they represent a supplementary danger because of the radiation emitted. The hazard can come from external irradiation and/or from contamination. The integrated equivalent dose is measured in Sievert (1 Sv = 1 Jkg^{-1}). The effects of exposition even to small doses of ionizing radiation are long-term effects (cancer) and genetic effects. Significant doses, of the order of many Sieverts, can produce in humans immediate serious consequences; massive doses (tens of Sieverts) within a short period of time can be fatal to humans. The maximum yearly reference dose for professionally exposed workers is 0.05 Sv. Some corresponding yearly intakes by inhalation are 2 10^4 Bq for Po^{210}, 2 10^2 Bq for Pu^{239}, and 2 10^6 Bq for I^{131}. A radioactive substance has an activity of 1 Bq (1 Bq = 2.7 10^{-11} Curie) when it has a disintegration per second. (See Chapter 2 for further details).

Table 3

**SOME POISONS AND PESTICIDES THAT CAN BE
INVOLVED IN EXPLOSION OR FIRE[37,38]**

Acrolein	Dinitrochlorobenzene	Nitroclorobenzene
Aldrin	Endrin	Nitrogen tetroxide
Allyl alcohol	EPN	Parathion
Ammonia	Ethylammine	Phenol (carbolic acid)
Arsenic compounds	Fluorine	Phosdrin
Beryllium	Guthion	Phosgene
Bromoacetone	Heptaclor	Procylamine
Carbon disulfate	Hydron cyanide	Rotenone
Chlorine	Hydrogen fluoride	Strychnine
Chloropicrin	Lindane	Systox
Cresol	Methylamines	TEPP
Cyanides	Methyl bromide	Tetraethyl lead
Cyanogen	Methylparathion	Thallium salts
Diborane	Mercury compounds	Thimet
Dieldrin	Nicotine	Thioden
Dimethyl sulfate	Nicotine sulfate	Toxaphene
Dimethyl sulfide	Nitrobenzene	Trithion
Dinitrobenzene		

D. Detection

The knowledge of the presence, nature, and amount of a dangerous substance leaking from an outside plant, if detected in time, allows preventive action to be taken. When an accident crosses the boundaries of the involved plant and there exists a well-organized emergency plan, much information and suggestions can come from the authority in charge. However, an internal detection system will be useful in completing information and constitutes a spare device to face outside danger.

Indicators can be general purposes or calibrated on a specific substance, portable or fixed, continuous or operating on samples.

Detection of a specific flammable or combustible gas can be furnished by the particular properties of the gas such as diffusion or thermal conductivity, refractive index, or density. Radioactive iodine-specific detectors are based on the energy of the particles emitted in the iodine atom disintegrations.

General purpose flammable or combustible detectors utilize solid-state sensors or catalytic combustion. In this last case, mixtures, even at very low concentration, burn on a filament of heated platinum, and a so-called catalytic combustion takes place. Maintenance, periodical calibration, operating tests of detectors, of the associated network and of the protection actions are needed. For instance, the platinum surface can lose its property if exposed to vapors of tetraethyl lead, silicones, and dusts.

E. Protection

In absence of a special filter-equipped ventilation system, any building immersed in an outside constant concentration, such as ammonia in air, in short time will have its inside concentration C_i near to the external one C_e. If λ is the number of air changes per hour in the building, the inside concentration after t hours will be

$$C_i = C_e(1 - e^{-\lambda t})$$

Therefore, 1 h after the immersion, C_i will be 0.63 times the external concentration and another hour later it will be 0.86 C_e, if λ is 1.

For a given release, the rate of the release assumes importance in determining the arising dangerous situation. If the rate is low, due to the in-leakage, the protection offered from the building soon will be ineffective. In a quantitative calculation chlorine releases of ~ 20 T at different rates (6 s, 600 s, 6 h) have been considered.[27] The weather conditions have been fixed class D and windspeed of 5 m/s. For each case, the mean concentration arising in the building due to the passage of the cloud has been calculated. Assuming that as soon as the cloud has passed the occupants renew contaminated air with fresh air (or leave the building) the lethal distance from the source is less than 200 m when the release is almost instantaneous, less than 500 m in 600 s release and approximately 600 m in 6 h release.

Filters, fans, and associated equipment should be located in a room cut off from the rest of installation and protected against fires by an automatic sprinkler system. Where the ducts pass through fire barriers, or fire walls, adequate firestopping must be provided to seal the space between the edges of the opening and the duct walls. The fresh-air intakes should be located as low as possible (since smoke generally rises) and taking into account all precautions to reduce the exposition to sparks or to external dangerous substances. Generally speaking, two types of air cleaning items can be adopted: filters and electronic air cleaners. Their scope is to remove any dust or particulate from the air stream. The filters classified in class I do not contribute fuel when attacked by flame and produce only small amounts of smoke. Fibrous filters can be dry or of the viscous impingement type; in the latter the fibers are coated with an adhesive which traps any foreign material.

When pressure drops due to entrained particles trapped and collected on the filter reach about 1 cm of water, the filter is removed, cleaned, and reinstalled or is replaced by a new one. A reference filter operational efficiency is of 99.97% with 0.3-μm particles.

Electronic air cleaners are based on the principle of electrostatic precipitation. Particles pass through adequate electrostatic fields and are collected on filters or on charged plates by polarization.

If the contaminant agents are iodine aerosols, special filters must be provided. Iodine isotopes are the most representative fission products released from nuclear reactors in case of accidents involving nuclear fuel and the reactor physical barrier integrity. The iodine isotopes inventory in a nuclear reactor after 1000-d burn-up is about 10^{16} Bq/thermal MW.[28] A typical recent generation electric power station hence will contain some 10^{19} Bq of iodine isotopes (of which more than one tenth is I^{131}) after 1000-d burn-up. Only a fraction of these isotopes will leave the fuel due to plate out and a still smaller fraction will be released outside the reactor containment because of filter retention. According to a recent German reactor safety analysis, the total I^{131} release following the worst considered accident in a 1300-MWe power station should be of about 2.3 10^{16} Bq;[29] another typical radioisotope, Cs^{137}, will accompany it in the amount of about 1.7 10^{15} Bq. However, the Chernobyl accident has shown that more important releases are possible; of course, each accidental scenario is linked to design criteria and regulatories rules concerned. Retention of iodine isotopes coming from an external close nuclear plant requests the installation of special charcoal filters; their efficiency for metallic iodine is about 99.99%; it is less for iodine compounds.

Depending on the type of danger and on the kind of duties of each one, people must be adequately protected to face internal implications and to save their health. The protection devices have to be of an approved type, tested and inspected at suitable frequencies, kept in proximity of potential needs, and stored in good state of preservation.

At least the protective equipment must include respiratory protection devices. Other equipment is suitable such as helmets, hoods to protect head, neck and ears, face shields to protect face and eyes, protective coats, gloves, and boots.

The human respiratory system can be protected with masks or with a self-contained breathing apparatus. Masks are filter equipped, each filter being specific for the substance (or group of substances) from which protection is needed. Self-contained breathing apparatuses are available in two types:

Table 4
VARIOUS SABOTAGE TARGETS: COMPARISON

	Success probability		
	Low	**Medium**	**High**
High			
	Nuclear weapons Warfare chemicals	Dam	Water supply Food supply Public gathering
Medium			
	Nuclear power reactor Munition depot	Public building Bridge Tunnel Airport and aircraft Explosives	Railroad yard and trains Docks and ships Toxic chemicals Petroleum and natural gas
Low			
	Military base	Bank Fossil-fuel power plant	Communications Power transmission

From Sandia Laboratories, *Nuclear Safety*, 665, 1976.

1. Open circuit apparatus. Exalation is vented to atmosphere and not rebreathed. The pressure inside the facepiece, in relation to the near environment must be positive always, during both phases of inhalation and exalation.
2. Closed circuit apparatus. Exalation is rebreathed by the wearer after the oxygen concentration has been restored and the carbon dioxide removed.

Both masks but especially self-contained breathing apparatuses should be used after appropriate training only.

III. PHYSICAL PROTECTION

Physical protection against sabotage and against diversion of dangerous substances is a problem of growing importance.

Such actions can be accomplished from people working inside a high-risk plant, from outsider saboteurs, and from outsiders in collusion with insiders. International and national organizations have been concerned in preparing general guides for defined plants and materials.[30-32]

The probability that a sabotage is attempted and when attempted is successful should be calculated. The consequences of successful sabotage should be evaluated too. Table 4 shows a qualitative ranking of some sabotage targets. The greatest risk (greatest probability, greatest consequences) is presented by activities in the upper right corner.

To face such a kind of danger, any high-risk plant should have a physical protection, according to local particular sociopolitical conditions. A physical protection system is based on:

1. Detection equipment to signal any undue act from insiders and to signal any intrusion or any intrusion attempt from outsiders

2. Protective structures around vital areas, having the double function of shielding the most important equipment and delay item to allow response force intervention
3. Response force to block and neutralize saboteurs' actions

Hence, prevention against sabotage is based on plant design, presence of security force, and administrative controls. Every effort should be made to reduce the time in detecting saboteurs' presence and to delay their action. Each additional second gained will provide additional time for response force and will improve the probability of success.

Plant design will provide the identification of so-called "vital areas" needing particular care and the identification paths that can be critical from a defense point of view. Vital areas should be isolated from the less important sections of the plant. Examples of vital areas can be control rooms; storage areas of flammables, explosives, poisons, radioactive materials, some chemical reactors and related equipment, emergency cooling systems, and electrical emergency systems. Vital areas shall be protected and enclosed by physical barriers. These barriers should be constructed in such a way that their integrity is not easily compromised from outside, e.g., when use of small fire arms is made. Therefore, the barrier should be able to deter intrusion of unauthorized persons. An empty area shall be left near two sides of the barrier to allow clear sight and surveillance. A succession of obstacles and barriers, with associated detection and intervention system, will be very effective.

Typical antitheft detectors are placed so that any intrusion attempt can be detected and registered. When this is the case, a few well-determined alarms can trip some safety process actions. Microwaves, sonic and ultrasonic sensors, television cameras, optical fibers, and vibration and pressure devices are all usually adopted.[33] A reliability analysis will check that failure probability is very low. The probability of spurious alarms must be low too.

However any defense criterion is worthless if an inside specialist having free access (due to his job) to safety system logics, deliberately tampers interlocks and leaves the process partially or totally unprotected. Therefore, an investigation to positions allowing access to some particular areas should be made. Investigation will concern true identity, character, education, employement credit, criminal history, and military service. Moreover, people allocated in these positions should be submitted to periodical psychological assessment and to continuous observation.

Active response is the responsibility of especially equipped and continually trained security force. Guards, watchmen, and patrol watchmen may be uniformed and armed according to particular regulations.[34] Duties must be clearly assigned and submitted to periodical revision.

Access to vital areas will be allowed only to personnel having need to enter these areas; access to protected areas will be limited. Authorization will be needed for access to vital and protected areas. People authorized to enter such areas without escort shall have tamper-resistant badges. Moreover, they should be personally recognized by guards and should be checked and accepted by particular devices controlling some defined physical features (fingerprints, etc.). Vehicle access must be regulated. Unannounced and/or regular search of personnel and of vehicles must be performed.

All tactics that saboteurs could adopt must be assumed and examined. On the other hand, the safeguard effectiveness of the plant to be protected, of the defense installations, and of the security force has to be verified. Then, simulation models can be very helpful and, in fact, models using Montecarlo, Markov, and event-tree techniques have been prepared and applied.[35]

In these models, the plant characteristics are represented, the potential performances of installations, of Security Force, and of saboteurs are defined, the several possible routes to reach the objectives are foreseen, the interposed defenses are considered, the different reactions are taken into consideration, including hypothetical fightings between attackers and response force[36] and the system sensitivity to fundamental parameters variations is evaluated.

In this way the simulation analysis allows to discover weak points and permits to take the needed corrective actions.

REFERENCES

1. **Strehlow, R. A.,** Unconfined vapor-cloud explosion. An overview, 14th Symp. Int. Combustion, The Combustion Institute, 1973, 1189.
2. **Kletz, T. A.,** Unconfined vapor cloud explosions, *Loss Prevention*, 50, 1977.
3. **Geiger, W.,** Generation and propagation of pressure waves due to unconfined chemical explosion and their impact on nuclear power plant structures, *Nucl. Eng. Design*, 198, 1974.
4. **Lannoy, A.,** Analyse des explosions accidentelles réelles, Revue générale de thermique, 209, 1982.
5. **Briscoe, F.,** A review of current information on the causes and effects of explosions of unconfined vapor clouds, Canvey, An investigation of potential hazards from the operations in the Canvey Island/Thurrock area, Her Majesty's Stationery Office, 1978.
6. **Wiedermann, A. H., Eichler, T. V., and Kot, C. A.,** Air blast effects on nuclear power plants from vapor cloud explosions, 6th Int. Conf. Structural Mechanics in Reactor Technology, Paris, 1981.
7. **Hendricks, S. and Lannoy, A.,** Réflections sur la modélisation de phénomènes d'explosions de mélanges air-hydrocarbone en milieu libre, Trans. 7th Int. Conf. Structural Mechanics in Reactor Technology, Vol J, Chicago, 1983.
8. **Rasbash, D. J.,** Review of explosions and fire hazards of liquefied petroleum gas, *Fire Safety J.,* 1979.
9. **Marshall, V. C.,** *The Bulk Storage and Handling of Flammable Gases and Liquids, An Oyez Intelligence Report,* Oyez, London, 1980.
10. **High, R. W.,** The Saturn fireball, *Ann. N.Y. Acad. Sci.,* 441, 1968.
11. **Marshall, V. C.,** Unconfined-vapor-cloud explosions, *Chem. Eng.,* 149, 1982.
12. **Kletz, T. A.,** Some questions raised by Flixborough, AIChE Loss Prevention Symposium, Houston, 1975.
13. **White, C. S.,** The scope of blast and shock biology and problem areas in relating physical and biological parameters, *Ann. N.Y. Acad. Sci.,* 89, 1968.
14. **Clemedson, C. J., Hellstroem, G., and Lindgren, S.,** The relative tolerance of the head, thorax and abdomen to blunt trauma, *Ann. N.Y. Acad. Sci.,* 187, 1968.
15. **Bowen, I. G., Fletcher, E. R., Richmond, D. R., Hirsch, F. G., and White, C. S.,** Biophysical mechanisms and scaling procedures applicable in assessing responses of the thorax energized by air-blast overpressures or by nonpenetrating missiles, *Ann. N.Y. Acad. Sci.,* 122, 1968.
16. **Richmond, D. R., Damon, E. G., Fletcher, E. R., Bowen, I. G., and White, C. S.,** The relationship between selected blast-wave parameters and the response of mammals exposed to air blast, *Ann. N.Y. Acad. Sci.,* 103, 1968.
17. **Hirsch, F. G.,** Effects of overpressure on the ear. A review, *Ann. N.Y. Acad. Sci.,* 147, 1968.
18. **Sperrazza, J. and Kokinakis, W.,** Ballistic limits of tissue and clothing, *Ann. N.Y. Acad. Sci.,* 163, 1968.
19. **Baker, W. E., Cox, P.A., Westine, P. S., Kulesz, J. J., and Strehlow, R. A.,** A short course on explosion hazards evaluation, Southwest Research Institute San Antonio, Houston, 1978.
20. **Jarrett, D. E.,** Derivation of the British explosives safety distances, *Ann. N.Y. Acad. Sci.,* 18, 1968.
21. **Praz, M.,** *Structural Dynamic: Theory and Computation,* Van Nostrand Reinhold, New York, 1980.
22. **Westine, P. S.,** R-W plane analysis for vulnerability of targets to air blast. V, *Shock Vibration Bull.,* Bulletin 42, 173, 1972.
23. **Baker, W. E., Westine, P. S., and Cox, P. A.,** Methods for prediction of damage to structures from accidental explosions, Proc 2nd Int. Symp. Loss Prevention, Heidelberg, 1977.
24. **Forbes, D. J.,** Design of blast-resistant buildings in petroleum and chemical plants, in *Safety and Accident Prevention in Chemical Operations,* Fawcett, H. H. and Wood, W. S., Eds., John Wiley & Sons, New York, 1982.
25. **Farrar, D. G., Hortzell, G. E., Blanh, T. L., and Galsten, W. A.,** Development of a Protocol for the assessment of the toxicity of combustion products resulting from the burning of cellular plastics, Report UTEC 79/130, University of Utah, Salt Lake City, 1979.
26. **Petajan, J. H., Voorhees, K. J., Packham, S. C., Baldwin, R. C., Einhorn, I. N., Grunnet, M. L., Dinger, B. G., and Birkey, M. M.,** Extreme toxicity from combustion products of fire-retarded polyurethane foam, *Science,* no. 187, 1975.
27. **Beattie, J. R.,** A quantitative Study of Factors Tending to Reduce the Hazards from Airbone Toxic Clouds, Appendix 3, An Investigation of Potential Hazards from Operations in the Canvey Island/Thurrock area, Her Majesty's Stationery Office, 1978.

28. **Kelly, J. L., Reynolds, A. B., and McGown, M. E.,** Temperature dependence of fission product release rates, *Nucl. Sci. Eng.* 88, 184, 1984.
29. **Hassmann, K. and Hosemann, J. P.,** Consequences of degraded core accidents, *Nucl. Eng. Design,* 285, 1984.
30. The Physical Protection of Nuclear Material, IAEA-INFCIRC/225/Rev. 1, International Atomic Energy Agency, Vienna, June, 1977.
31. Convention on the Physical Protection of Nuclear Material, Legal series no. 12, International Atomic Energy Agency, Vienna, 1982.
32. Industrial security for nuclear power Plants, ANSI no. 18.17,
33. USNRC, Perimeter Intrusion Alarm Systems, Regulatory Guide 5.44, U.S. Nuclear Regulatory Commission, Washington, D.C.
34. USNRC, Training, Equipping and Qualifying of Guards and Watchmen, Regulatory Guide 5.20, U.S. Nuclear Regulatory Commission, Washington, D.C.
35. **Polito, J.,** The Safeguards Network Analysis procedure: A User's Manual, NUREG/CR-3423, SAND83-7123, 1983.
36. **Engi, D. and Harlan, C. P.,** Brief Adversary Threat Loss Estimator User's Guide, NUREG/CR-1432, SAND80-0952, 1981.
37. **Jones, R.,** Agricultural chemicals as a fire hazard, *Fireman,* 32(4), 1965.
38. NFPA Inspection Manual, Quincy, MA, 1982.
39. Sandia Laboratories, Safety and Security of Nuclear Power Reactors to Acts of Sabotages, *Nuclear Safety,* 665, 1976.

Chapter 5

ARTIFICIAL EVENTS: AREA EVENTS

Alberto Ferreli

TABLE OF CONTENTS

I. INTRODUCTION*

Area events are defined as all the possible effects consequent to the breakdown or the malfunction of whatever systems are relevant to safety or to health protection. The term "area" identifies a particular zone of a plant (namely a nuclear plant) which contains systems or components that are relevant to safety and to health protection, in which the effects following the hypothesized incidental events can be circumscripted. Typical area events are

1.   Flooding
2.   Pipe whip and jet forces
3.   Pressure and thermal loads
4.   Missiles
5.   Dropping of heavy loads
6.   Fire and explosions

Since their effects may represent a "common cause of failure"** for all the systems and components which are located in the same area, they may represent a serious detriment for the safety of the plant. They are in particular dealt with systematically in nuclear applications.

In order to cope with these events, "safety-related"*** functions are normally performed by redundant components located in different areas. When this is not easily achievable, safety-related systems are designed to carry out their functions under the most severe environmental conditions associated with such events.

Among the area events, the accidental release of toxic gases is to be quoted, since it could cause injuries to operating personnel or prevent the entrance to important areas, with possible loss of plant control.

II. FLOODING

A rupture in a "high" or "moderate energy" system† could result in water or other liquids falling to the floor and draining to adjacent regions, with possible adverse consequences in the flooding areas, such as

- Submersion of components
- Electric circuitory damaging
- Rise of additional hydrostatic loads
- Possible contamination of the area
- Loss of accessibility

Protection against flooding includes:

- Provisions for appropriate detection and isolation of a flooding, so to minimize consequences

---

* See note 1 in Chapter 2.

** The "common cause of failure" is defined as a failure of a number of devices or components to perform their functions as a result of a single specific event or cause.

***See note 2 in Chapter 2.

† A "high energy" system is commonly defined as a system either containing a fluid whose maximum operating temperature exceeds the boiling point or whose maximum operating pressure is well above the atmospheric pressure. A "moderate energy" system is defined as a system whose operating parameters are below the above specified values.

- Provisions for sufficient drainage capacity
- Location of vulnerable parts of components above the estimated water levels, considering also potential splashing or wave action

The maximum flooding level should take into consideration potential sources of liquids located in each area. In the design of nuclear power plants it has become a common practice to divide the plant into flooding areas capable of preventing the spread of water to other areas. In particular, care is paid in avoiding communications among the drainage systems of different areas; when possible, headers serving an area are channeled to their own sump; in other cases drainages are equipped with inspectable and testable check valves to prevent a backflow from the flooding in other areas.

The entire drainage system should be designed to withstand the design basis earthquake, if consequences of its damage are considered unacceptable. For each single flooding area, starting from the assumption that a flooding occurs, an evaluation is carried out to verify sufficient availability of systems in order to ensure the ''safe shutdown''* of the plant.

In case of dispersion of hazardous liquids, provisions are taken to assure control of the contamination spread within and outside the plant, transfer and storage of the liquids in the waste disposal system, and re-establishment of operating conditions in the area involved without undue risk for the workers.

III. PIPE WHIP, JET FORCES

One of the effects associated with a break in a pipe containing a high-energy fluid — such as steam, flashing water, or pressurized gas — is that known as ''pipe whip'', which consists in the uncontrolled motion of the two pipe ends, dynamically excited by reaction forces induced from the rapid blowdown of the fluid through the break.

Structures and components, within the range of the free ends of the broken pipe, could be damaged as the result of the impact.

Also associated to high energy pipe breaks are jet forces arising from the impingement of the jet on nearby structures and components.

A. Piping Failure Hypotheses

Evaluation of piping failure effects are based upon the assumption of various types of failure, in relation to the pipe size and characteristics, the energy of the fluid, and other operating conditions. Failure locations are commonly assumed at

- Terminal ends of the pipes
- Intermediate locations chosen at welding seams
- Other locations where the stress intensity factor is above defined values

Regarding the break dimensions, the following hypotheses are commonly used in the design of nuclear power plants:

1. Instantaneous circumferential break in high-energy systems
2. Slowly opening circumferential crack in moderate energy systems
3. A longitudinal crack in high or moderate energy systems.

Specifically for the protection against dynamic effects, a break smaller than the pipe cross-

---

* ''Safe shutdown'' is defined as plant condition where the plant is shut down and all the auxiliary functions needed to maintain the plant in a safe steady condition are performed. Containment of dangerous substances is included.

section or "no-break" criteria may be assumed where specific provisions have been taken to ensure that adequate safety margins and high quality requirements for materials, design, construction, and inspection are met.

B. Protection Against Pipe Whip and Jet Forces

To cope with dynamic effects of piping ruptures the following provisions are adopted:

1. Where possible, systems containing high-energy fluids are housed in different areas from those where the safety-related systems are located
2. Redundant components of safety-related systems are segregated from each other

Where it is not possible to achieve protection by means of the above-mentioned measures, such a protection is ensured through the adoption of

- Adequate distances
- Pipe whip restraints capable of absorbing the kinetic energy associated with the broken pipe and limiting its motion
- Jet deflectors and jet impingement shieldings to protect safety equipment from jet forces

The adequacy of the measures adopted are based upon specific safety analyses; potential pipe motions at postulated rupture locations are evaluated to verify that safety equipment is not damaged by the effects of pipe whips and jet forces; analyses include effects of jet loading, fluid temperature, and moisture on the target impinged upon.

Simplified methods and models to evaluate fluid thrust forces, jet geometry, and jet impingement effects are presented in Reference 4.

IV. PRESSURE AND THERMAL LOADS

Other effects associated with the discharge of high-energy fluid in a compartment are the pressure and temperature loads following the compartment pressurization. It is normally a short-term event, occurring in a few seconds or less. Pressurization may damage structural elements or other equipment inside the compartment, because of the combined effects of high temperature and differential pressures set up in the area. Protection against compartment pressurization is assured by:

1. Appropriate leak detection and fast isolation of the broken system, in order to minimize the quantity of fluid released
2. Relief devices — such as blowout panels, opening through barriers, walls, or floors in narrow areas — where high-energy fluids are located

V. MISSILES

Pressurized vessels, piping and valves, high-speed rotating equipment, and storage of compressed or explosive gases, have a significant potential for missile generation in case of a failure.

A. Components Containing High-Energy Fluids

All pressurized components, such as vessels, pipe fittings, and valve bodies, present significant potential for missiles in case of rupture. They may fail in different modes, depending on the shape, welds position, material characteristics, operating conditions, etc. Because of the severe consequences generally associated with the failure of high-pressure components — especially pressure vessels — which are not limited to the missile generation but involve the whole safety of the plant, the general preventive measures adopted assure in many cases a sufficient protection against the specific risk of dynamic effects.

Common practice in reducing pressure retaining components failure probability consists of

1. Definition of stringent design limits, based on accurate analyses of static and transient loads and application of safety factors
2. Control of material properties and fabrication in order to identify preserve existing flaws
3. Periodic in-service inspection to detect flaws which may become critical during operation
4. Surveillance systems, such as leak detection or loose parts detection systems, to monitor conditions which may give indication of incipient failure.

B. High-Speed Rotating Equipment

Rotating machinery — such as steam turbines, pumps, and their motors and flywheels — are very common components in industrial plants. Rotating parts may reach in operation considerable energy which, in case of failure, may be converted in translational kinetic energy of rotor fragments.

Potential for producing missiles may arise from overspeed due to sudden reduction in the load or sudden increase in the input energy; if the speed gets too high, the stress resulting from centrifugal forces in the rotor can destroy the rotating assembly. Therefore, rotating machines have speed control features and provisions for preventing destructive overspeed such as governors, clutches, brakes, and a combination of instrumentation, control, and valving systems.

These provisions do not exhaust the measures that should be taken to reduce the failure probability to an acceptable level; design, manufacturing, operational quality assurance, and in-service inspection are also requested to avoid the possibility that a flaw in the rotor may cause its failure also at or below normal running speed.

Other provisions, different from the preventive ones, are represented by physical separations among the potential missile sources and the components to be protected. In many cases also the plant layout plays an important role, by locating safety-related parts out of the trajectory of possible missiles.

In nuclear power plants, for instance, the turbo generator axis is frequently placed in radial direction with respect to the reactor.

C. Storage of Compressed Gases

Compressed gases are extensively used in industrial application; for instance, oxygen is used in different processes and in metal-cutting and welding equipment, nitrogen in establishing an inert atmosphere, or to fill accumulators, carbon dioxide and halon in fire extinguishers. Hydrogen is used in process and in large electrical generators to reduce windage losses, or as a cover gas in water tanks to inhibit corrosion. Sulfur hexafluoride, because of its electric resistivity, is used as an isolating atmosphere in electrical equipment, such as switchyard circuit breakers.

Compressed gases are hazardous because of their stored energy, which could be released in case of equipment failure.

Gas storage tanks and receiving areas shall be isolated from other equipment. Gas containers shall have adequate design margins and shall be equipped with reliable relief devices.

High-pressure gas lines and containers should be identified so to be easily recognizable.

Of particular concern are portable cylinders or "gas bottles" because they are used throughout the industrial activities and serious accidents involving cylinders have occurred. They should be stored in a proper environment, periodically inspected, and handled according to well-defined procedures; in particular their presence in areas containing safety-related equipment should be forbidden, unless special procedures are developed to protect the cylinders and to prevent their failure.

D. Barrier Design

The response of a barrier to a missile impact depends on many factors such as the shape of the missile, its kinetic energy, and the barrier characteristics.

Sufficient thickness of material should be provided to prevent perforation; spalling and scabbing should also be avoided in designing concrete structures or barriers.

Empirical equations to estimate missile penetration and to determine the required barrier thickness are reported in Reference 15 for concrete structures, and in References 16 and 17 for steel plates.

The resistance of multi-element missile barriers may be evaluated assuming the residual velocity of the missile perforating the first element as the striking velocity for the next element.[18] To evaluate the overall response of the structure to the impact, besides the demostration that the missile will not penetrate the barrier, an equivalent static load concentrated at the impact area should be assumed. A method for such analysis is presented in Reference 19.

VI. DROPPING OF HEAVY LOADS

Heavy loads may be handled in several plant areas and, if they were to drop in certain locations, they might impact safety-related components. Adequate measures are established in order to

1. Define safe load paths so to minimize the potential for heavy loads to impact, if dropped, safety-related equipment
2. Develop procedures to cover load handling operations in proximity of safety-related equipment
3. Train and qualify crane operators

Where unacceptable results are expected from the effects of postulated load drop, further protections are provided by employing single-failure-proof handling systems.

VII. FIRE AND EXPLOSIONS

The plant and their structures, systems, and components shall be designed in such a manner as to reduce to the minimum — in respect of other safety requirements — the probability of fire and explosions and their effects on safety-related equipment.

To this end noncombustible, flame-retardant, and heat-resistant materials are employed as extensively as possible. Moreover, redundant parts of safety-related systems are separated by suitable distance or by fire-resistent barriers. Detection and extinction systems of adequate capacity and reliability are provided for. Fire-fighting systems are designed and positioned so as to ensure that their failure or involuntary or unnecessary activation shall not substantially prejudice the functionality of safety-related structures, systems, and components.

Special precautions shall be taken to avoid explosions.

Lines containing explosive gases should be easily recognizable.

Hydrogen lines, in areas housing safety-related equipment, should be either seismically designed or sleeved such that the outer pipe be directly vented to the outside, or should be equipped with excess flow valves.

Areas where it is possible for explosive gases to accumulate, should be adequately monitored and ventilated. In particular, consideration should be given to possible chemical reactions which may generate explosive gases, such as the hydrogen produced inside the battery rooms.

For a further analysis, see Chapter 3.

VIII. RELEASE OF ASPHYXIANT AND TOXIC SUBSTANCES

Since most of the compressed gases are asphyxiant or toxic they may represent a hazard first for the operating personnel.

This circumstance plays an important role for the general safety of the plant, beyond the specific aspect of the workers protection. In fact the operators could be prevented from performing safety actions needed in accident conditions.

Specific protection is required against hazardous chemicals that may be discharged as a result of equipment failures or operator errors.

Among the different chemicals stored at the plant, chlorine — used in many nuclear power plants for water treatment in the circulating water system and in other auxiliary systems — is specifically recognized as a potentially hazardous chemical.

Maximum quantities of chlorine which can be stored at the plant and minimum distances from the control room are defined in many countries.

Risk of drifting clouds of toxic gases may arise also from hazardous industrial plants or transport routes in the neighborhood of the plant, and the extent of such a risk depends on the distances, quantities stored or transported, frequencies of the shipments, etc.

Protective provisions of the control room against these hazards include

1. Adoption of an emergency closed-circuit ventilation system
2. Redundancy and separation of fresh air inlets
3. High location of fresh air inlets
4. Limited rate of normal fresh air makeup
5. Low-leakage construction of the control room, with a limited air exchange rate
6. Quick response of the available chemical detectors located in fresh air inlets, which initiate complete closure of isolation dampers on detection

In some countries, such as Germany, Italy, and Belgium, it is also required, as a further precaution, that, in case of unavailability of the control room, a special emergency system automatically shuts down the plant and keeps it in a controlled state without any manual intervention.

REFERENCES

1. Safety Implications Associated With In-Plant Pressurized Gas Storage and Distribution Systems in Nuclear Power Plants NUREG/CR-3551, U.S. Nuclear Regulatory Commission, Washington, D.C., April 1985.
2. Protection Against Internally Generated Missiles and Their Secondary Effects in Nuclear Power Plants, Safety Series No. 50-SG-D4, International Atomic Energy Agency, Vienna, 1980.
3. Reactor Coolant and Associated Systems in Nuclear Power Plants, Safety Series No. 50-SG-D13, International Atomic Energy Agency, Vienna, 1986.
4. Design Basis for Protection of Light Water Nuclear Power Plants against effects of Postulated Pipe Rupture, ANSI/ANS 59.2.1980, American Nuclear Society, 1980.

5. Assumption for Evaluating the Habitability of a Nuclear Power Plant's Control Room During a Postulated Hazardous Chemical Release, Regulatory Guide 1.78, U.S. Nuclear Regulatory Commission, Washington, D.C., June, 1974.

6. Protection of Nuclear Plant Control Room Operators Against an Accidental Chlorine Release, Regulatory Guide 1.95, Rev. 1, U.S. Nuclear Regulatory Commission, Washington, D.C., January 1977.

7. Protection Against Low-Trajectory Turbine Missiles, Regulatory Guide 1.115, U.S. Nuclear Regulatory Commission, Washington, D.C., July 1977.

8. Internally Generated Missiles (Outside Containment), Standard Review Plan 3.5.1.1, Rev. 2, U.S. Nuclear Regulatory Commission, Washington, D.C., July 1981.

9. Internally Generated Missiles (Inside Containment), Standard Review Plan 3.5.1.2, Rev. 2, U.S. Nuclear Regulatory Commission, Washington, D.C., July 1981.

10. Turbine Missiles, Standard Review Plan 3.5.1.3, Rev. 2, U.S. Nuclear Regulatory Commission, Washington, D.C., July 1981.

11. Barrier Design Procedures, Standard Review Plan 3.5.3, Rev. 1, U.S. Nuclear Regulatory Commission, Washington, D.C., July 1981.

12. Plant Design for Protection Against Postulated Piping Failures in Fluid Systems Outside Containment, Standard Review Plan 3.6.1, Rev. 1, U.S. Nuclear Regulatory Commission, Washington, D.C., July 1981.

13. Determination of Rupture Locations and Dynamic Effects Associated with the Postulated Rupture of Piping, Standard Review Plan 3.6.2, Rev. 1, U.S. Nuclear Regulatory Commission, Washington, D.C., July 1981.

14. RSK Leitlinien für Druckwasserreaktoren, GRS 5/82, Gesellschaft für Reaktorsicherheit mbH, Germany, 3. Ausgabe October, 1981.

15. **Kennedy, R. P.**, A Review of Procedures for the Analysis and Design of Concrete Structures to Resist Missile Impact Effects, Holmes and Narver Inc., Orange, CA, September 1975.

16. **Cottrell, W. B. and Savolainen, A. W.**, U.S. Reactor Containment Technology, ORNL-NSIC-5, Vol. 1, Oak Ridge National Laboratory, Oak Ridge, TN, 1965, chap. 6.

17. **Russell, C. R.**, *Reactor Safeguards*, MacMillan, New York, 1962.

18. **Recht, R. F. and Ipson, T. W.**, Ballistic perforation dynamics, *J. Appl. Mech. Trans. Am. Soc. Mech. Eng.*, 30, Series E(3), 1963.

19. **Williamson, R. A. and Alvy, R. R.**, Impact Effect of Fragments Striking Structural Elements, Holmes and Narver, Inc., Orange, CA, Revised, November 1973.

Chapter 6

ARTIFICIAL EVENTS: HUMAN FACTORS

Ivo Tripputi

TABLE OF CONTENTS

I. INTRODUCTION

The human factor discipline is concerned with designing machines, operations, and work environment to match human capabilities and limitations. It has become increasingly important over the past years along with the increase in the size of conventional and nuclear power plants (due to scale economic reasons) and with the marked tendency to centralize controls and concentrate critical decision-making processes. All this puts the achievement of operation and safety goals in the hands of a small group of plant-operating staff. Many efforts have been made to facilitate the job and training of operators. However, human beings may fail as well as equipment; it is clear that even the use of microcomputers and robotics will not eliminate the pervasive influence of human errors. In the foreseeable future, many fundamental tasks will continue to be assigned to humans. For instance, control room operators are likely to remain in supervisory roles to deal with unexpected contingencies and act when necessary. Further, the safety and control system reliability will always depend on human errors. In this chapter, we will discuss all human-plant interactions in the plant operation. They include normal and emergency control room activities, equipment maintenance, and testing and calibration. In spite of their importance, human errors in design, manufacturing, construction, and installation are not covered because they would require a completely different treatment and usually are not technical but organizational issues. A recently published Institute of Nuclear Power Operations (INPO) work [1] has identified 111 significant events that occurred in U.S. nuclear power plants in 1985. They include actual or potential losses of safety functions, severe plant transients, major economic impact, occupational radiation exposure, or release of radiation. They have been analyzed to identify the root causes; it has turned out that 46% of them was caused by human performance problems in the plant operation. PRA results confirm that at least 50% of the total risk derives from human errors. On a general level, all major nuclear power plant accidents (and surely most of other types of industrial plant accidents) are negatively affected by human errors. "Perfect" operators would have avoided serious consequences and accidents in most cases.

A clear example is the Three Mile Island (TMI) accident which occurred in 1979 in a nuclear power plant in Pennsylvania. A large number of commissions studied the event in all its aspects and identified numerous contributing causes (poor design, licensing, and organization). However, human errors were indicated as the major direct cause. The operators voluntarily stopped the plant safeguards (that had already been correctly activated by the protection system) and were unable to diagnose the cause of the event and the core conditions for approximately 12 h after the accident took place. The President Commission Report concluded that " . . . the fundamental problems relate to people rather than equipment . . . ." The major cause of that accident was identified in the improper action by operators along with a number of other reasons:

1. The control room was not correctly designed
2. Over 100 alarms went off in the early stage of the accident
3. The arrangement of controls and indicators was improper
4. Several instruments were out of range during the accident

However, only after the TMI accident took place, extensive efforts were devoted to substantially improve the "man-machine" interaction in nuclear power plants.

The most tragic accident in a civilian nuclear power plant recently occurred in Chernobyl, involving a Soviet water-cooled, graphite-moderated plant. The reactor design had a positive reactivity coefficient being, therefore, intrinsically instable and for this a large number of computers and automatic controls had been installed. Soviet reports state that such plants operate very well in steady-state conditions. Quite probably, the operators overrated the safety margins of the plant. On April 26, 1986 the operators voluntarily defeated several automatic safety systems and neglected a number of operational limits established in the technical specifications to carry out a test on the power supply system as quickly and "efficiently" as possible.

In a few seconds, the plant went out of their control and a hell of radiations and flames lit the night.

Most features of the Chernobyl plant are quite different from those of Western countries. Nevertheless, that tragedy confirms the importance of human factors and plant mismanagement to ensure the plant safety.

From the above considerations, it could seem that equipment is always reliable while the human factor is a true safety limit in industrial plants. Nevertheless, although it is impossible to design a man-machine system void of human error or totally eliminate hardware failures, it is possible to radically reduce their negative influence on plant safety by properly using human factor discipline criteria. The human factor analysis comprises principles derived from other disciplines such as psychology, physiology, sociology, instrumentation design, control and workspace design, personnel selection, and training.

The need for an official discipline of this sort was first felt during the second World War. At the beginning of the war, the engineering attempts at reducing human errors in complex military systems were hardly successful, at best. Therefore, it was accepted that a discipline studying human components (in the same way that other disciplines study equipment components) should be followed to achieve an acceptable system reliability. For many years, the human factor expertise was almost exclusively applied to military system developments. With the advent of the space age, they started to consider exceptional requirements in terms of system reliability. Over the past decade, the application of human factors has been of benefit to numerous fields: information systems, consumer products, medical facilities, to name but a few.

As far as nuclear power plants are concerned, the interest in large-scale human factor development dates back to the Wash-1400[2] publication. It demonstrated the risk increase potentially attributable to discrepancies in the design of control rooms, panels, and display/control instruments. However, as previously mentioned, only after the TMI accident has a decisive effort been undertaken.

In the following sections we will present some criteria on human error classification (Section II), the element which may influence the human error probability (Section III), the meaning of the so-called integrated human factor program in the design of industrial plants (Section IV), an overview on issues, problems, and solutions for control room design (Section V), plant maintainability (Section VI), the procedure used to correctly select and train operators (Section VII), guideline procedures to operate the plant (Section VIII), and some final highlights to quantitatively assess the human error influence on the overall plant risk and the effectiveness of human factor criteria within the plant (Section IX).

II. HUMAN ERROR CLASSIFICATION

Many studies on human reliability place a considerable emphasis on the classification of human errors; an effective taxonomy is necessary if one is attempting to build up a store of data on human errors. A meaningful classification gives a useful insight into the way in which errors are caused and can be prevented. Many different schemes have been proposed as reversible-irreversible, systematic-random, and omission-commission. We will mainly refer to Reference 3 in classification, even if also other types could be conveniently used to our purposes.

Either under normal or abnormal plant conditions, any action performed by the plant

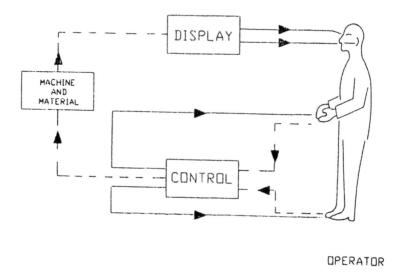

FIGURE 1.   The flow of information between operator and machine.

personnel is the final result of a complex physical and mental coordination which may be conveniently divided into a step-by-step process. This pattern is very important in the error classification. For instance, when referring to the control room operator response, his activities may be subdivided into the following overlapping categories:

1. Perception: to notice that some abnormal condition exists, e.g., that some alarms are sounding or blinking
2. Discrimination: to distinguish the signal characteristics (or a set of signals), e.g., the coolant level in a tank
3. Interpretation: to give a precise meaning to the signal discriminated, e.g., to realize that a low level of coolant in a tank means that there is a leak somewhere
4. Diagnosis: to determine the most likely cause(s) of the abnormal event
5. Decision making: to choose among different diagnoses and decide what to do
6. Action: to carry out the activities indicated by the diagnosis, operating rules, or written procedures

In the general model of the human component in the man-machine system (Figure 1), perception and discrimination tasks are internal inputs to the human processor, interpretation and decision-making are cognitive activities, while the actions carried out are the response. Other classifications summarize the above steps reducing them to two phases: (1) diagnosis and (2) action. A third phase may be added: recovery.

The latter scheme is generally used in the PRA analysis to quantify the system failure probability.

A different classification of the same errors, based on the timing of errors against the accident, may be presented. As to specific insights, we may distinguish the following: latent errors, dynamic errors, and recovery errors.

Latent errors are human errors made before the accident but discovered only afterwards. Dynamic errors are made in response to the accident. Recovery errors are made when one tries to restore the system or component operability or tries to correct the negative effects of a previous error. Furthermore, for the sake of completeness, it may be also convenient to classify incorrect human outputs as errors of omissions and commissions, including a more detailed breakdown as illustrated in Figure 2.

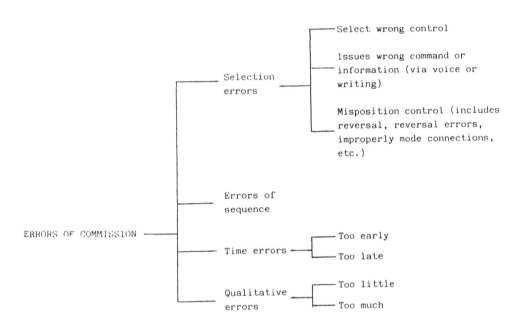

FIGURE 2. Error classification.

In each of the above-described steps the operators can make "errors". But what do we mean by human error? A human error is defined as any portion of a set of human actions exceeding some limit of accceptability. Thus, an error is merely an out-of-tolerance action, where the limit of the tolerable performance is defined by the system designer.

More specifically, one can make an error when he does something wrong, fails to do something he is supposed to do, or fails to do something in time (Table 1).

III. PERFORMANCE-SHAPING FACTORS

The Performance Shaping Factor (PSF) is any factor that influences the human performance. In this Section we will outline some qualitative aspects considering the human being as a system component, interacting with the other components by means of maintenance, calibration, testing, and normal operation and coping with those unusual events that place a system at risk.

We may divide PSF into three classes: external, internal, and stressors.

External PSF — It defines the work situation of operators, technicians, maintenance personnel, engineers, clerks and all those who take care of the reliability and safety of the plant. It includes three general categories:

Table 1

EXAMPLES OF ESTIMATED PROBABILITIES OF ERRORS OF COMMISSION IN OPERATING MANUAL CONTROLS

Item	Potential errors	HEP	EF
1	Inadvertent activation of a control	See text	
	Select wrong control on a panel from an array of similar-appearing controls[b]		
2	Identified by labels only	0.003	3
3	Arranged in well-delineated functional groups which are part of a well-defined mimic layout	0.001	3
		0.0005	10
4	Turn rotary control in wrong direction (for two-position switches, see item 8)		
5	When there is no violation of population stereotypes	0.0005	10
6	When design violates a strong populational stereotype and operating conditions are normal	0.05	5
7	When design violates a strong populational stereotype and operation is under high stress	0.5	5
8	Turn a two-position switch in wrong direction or leave it in the wrong setting	— [c]	
9	Set a rotary control in an incorrect setting (for two-position switches, see item 8)	0.001	10[d]
10	Failure to complete change of state of a component if switch must be held until change is completed	0.003	3
	Select wrong circuit breaker in a group of circuit breakers[e]		
11	Densely grouped and identified by labels only	0.005	3
12	In which the PSFs are more favorable (see text)	0.003	3
13	Improperly mate a connector (this includes failures to seat connectors completely and failure to test locking features of connectors for engagement)		

[a] The HEPs are for errors of commission only and do not include any errors of decision as to which controls to activate.

[b] If controls or circuit breakers are to be restored and are tagged, adjust the tabled HEPs.

[c] Divide HEPs for rotary controls (item 5 to 7) by 5 (use same EFs).

[d] This error is a function of the clarity with which indicator position can be determined: designs of switches knobs and their position indications vary greatly. For plant-specific analyses, an EF of 3 may be used.

[e] If controls or circuit breakers are to be restored and are tagged, adjust the tabled HEPs.

Adapted from Human Factors Guide for Nuclear Power Plant Control Room Development, EPRI NP-3659, December 1985.

1.   Situational characteristics: they are often plantwide in influence or cover many different jobs and tasks. They comprise architectural features, working environment quality, working hours and breaks, shift rotation and night work, availability/adequacy of special equipment/tools and supplies, manning parameters, organizational structure and external actions, rewards, recognitions, and benefits.

2.   Task and equipment characteristics: they are task specific in nature and include perceptual and motor requirements, control-display relationship, anticipatory and interpretation requirements, decision-making processes, complexity (information load), frequency and repetitiveness, task criticality, long and short-term memory, calculational requirements, feedback, dynamic vs. step-by-step activities, team structure, and man-machine interphase factors.

3.   Job and task instuctions: they include directives such as written and unwritten procedures, written or verbal communications, cautions and warnings, working methods, and plant policies.

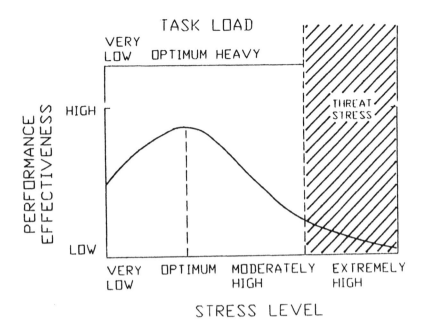

FIGURE 3.   Hypothetical relationship between performance and stress (based on Figure III 6-1 from WASH-1400) with task stress and threat stress division.

Internal PSF — It comprises both personal attributes as skill, ability, and attitude and factors related to previous education and training. Generally, the plant staff are properly selected and a margin of improvement in internal PSF is lower than in external PSF. However, improvement may be expected by means of accurate training and readiness towards emergencies, as further explained in Section VII.

Stressors — Stress is an internal PSF; however, it is handled differently than the other PSFs because of its special importance. We define stress as a bodily or mental tension and a stressor as any external or internal force causing stress. People can experience either psychological or physiological stresses. Often, stress is a combination of both. An adverse level of stress can arise when there is some mismatch between external and internal PSFs. For instance, if the perceptual requirement of a task puts too many demands to a worker, his performance will usually suffer because of an excessive task loading, a psychological stressor. On the other hand, without a sufficient task loading, the operator's performance might be degraded since stress is insufficient (mental or body tension) to keep him alert at work. This problem is as important as the first one. The operator's job has significantly changed over the last few years. As systems become increasingly automatic, the operator's role shifts from active participation and interaction with the plant to a system monitoring function. Therefore, his routine job is simplified but he may get bored or complacent. Then, when an emergency occurs, he is prone to commit errors. Figure 3 is a qualitative representation of the operator performance effectiveness related to task load and stress level.

To give a very rough quantification of the influence exercised by stressors, we may refer to the evaluation made in Reference 2; it leads to a conservative estimate of 0.2/0.3 as the average error rate for nuclear power plant personnel in a highly stressing situation such as a loss of coolant accident. This value has been derived from the observation of military aircraft crew error rates during in-flight emergencies and army recruits when exposed to simulated (but believed as real) attacks. In the first case, 0.16 average error rate was obtained while in the second one the rate was of approximately 0.3.

IV. HUMAN FACTOR PROGRAM IN DESIGN

As far as the system development is concerned, it is important to structure a plan for human factor program. The human factor objective in the program planning is to identify those development decisions that affect the personnel performance in the plant operation and to specify the efforts necessary to produce timely recommendations which take into account human capabilities. It reflects the awareness of the fact that it is possible to improve safety and productivity through the application of human behavior knowledge when development decisions affecting the personnel performance are to be made.

Some general guidelines to plan a human factor program include the identification of the following conceptual steps:

1. The plant staff are to be considered as a vital subsystem; their performance requirements should be defined and addressed in order to satisfy the plant overall requirements.
2. Interfaces should be identified, including environment, equipment, and documentation.
3. Development phases should be conceived as a framework including major design steps as conceptual, preliminary, detailed design, construction, and operation; the analysis will determine when to initiate human factor efforts, how to sequence them, which order should be followed to address interfaces, and how to determine the focus of the human factor program.
4. Determine the organizational position of human factors and needs and their coordination.

After solving basic conceptual issues, a specific program can be structured.

V. HUMAN FACTOR IN CONTROL ROOM DESIGN

The purpose of this section is not to provide detailed design criteria for control rooms, but rather to identify those issues which will require attention during the design that might influence the operator's performance as well as the logical process to be followed. Detailed criteria can be found in literature;[5] they cover information quality and quantity, information synthesis vs. elementary data, and ergonomics.

After structuring and approving a general human factor program, it is possible to undertake a specific analysis of functions, systems, and tasks as far as the control room is concerned. The human factor analysis relates known human behavior to ever more refined development decisions affecting the personnel performance. An effective human factor analysis organizes information on the plant to identify process and automatic control systems and their functional interrelationship, other than specifying detailed control functions. Afterwards, recommendations are prepared to allocate control functions either to automatic equipment or to the personnel subsystem.

After identifying the various kinds of task performed by the personnel, a staffing concept is established. Then, information and response requirements are progressively refined as actions, behaviors, and capabilities involved in the task performance are described and analyzed.

In the task analysis an important tool is represented by the availability of operation guidelines describing the action required by control room operators both in normal and emergency conditions (Section VIII). Once a firm analytical foundation has been provided, design and related evaluation efforts can be initiated, starting with the control room. Human factor efforts produce recommendations for the layout of the control room and the establishment of ambient environmental conditions to support the personnel performance. Ineffective control room design and major items (e.g., console, furniture, supervisory offices) will be arranged to facilitate the task performance. Consoles shall have a compact layout offering an easy access to display/control instruments. Console profile will be compatible with task requirements and permit instruments to be mounted on panels within the visual and functional reach of the personnel, taking into account the average physical characteristics of operators, which will be part of the selection criteria. Illumination for visual tasks is important, while distracting environmental effects (e.g., glare, noise, uncomfortable climatic conditions) shall be reduced or eliminated. In addition, other features promoting the personnel comfort, convenience, and safety will be considered, including furniture, lunch areas, restrooms, etc.

After making basic decisions on control room design, and panel design, related evaluation efforts will be initiated. The human factor efforts produce recommendations defining the type of instruments and their layout on panel surfaces available, and utilize labeling, demarcation, and coding techniques to support the task performance. The effective panel design should be compatible with what the staff thinks about the plant operation, sequence, frequency, and instrument priorities during tasks. In general, hierarchical, mimic, or partially mimic structures are extremely useful and reflect the functional dependencies among system components in the organization of instruments on panels. Displays and controls shall be consistently either integrated on common panel areas, segregated on vertical panel segments, or placed accordingly to a combined segmented-integrated approach. Common expectancies and visual scanning patterns will also be reflected in the display and control arrangements, which are labeled and demarcated to enable a rapid discrimination and recognition of instruments and groups of related instruments. Instruments should be coded to further facilitate the task performance, while controls should be located and protected to minimize potential accidental activation.

When the panel design efforts are under way, human factor efforts concerning alarm and announcing systems should be initiated. Such efforts help in establishing a coherent, consistent rationale to alert and inform the staff on different kinds of deviant plant conditions. Effective systems distinguish the nature of alarm conditions, according to established criteria, and either, using an alarm-associated code or an associated message, give basic information for the following action. The system is completely integrated into the plant control system; its design takes full advantage of the functional dependency knowledge within the plant operation and offers information compatible with human capabilities and limitations.

In general terms, any display provided with an attention-getting signal is said to be ''annunciated'' and is commonly called ''annunciator''. Annunciators of interest in nuclear power and most industrial plant control room are legend lights and automatic printout equipment. Other auditory alarms include high radiation, fire and security alarms, etc.

Usually, annunciated legend lights (or tiles) are mounted on panels above the vertical control boards, above the eye level. There can be 450 to 750 tiles per nuclear reactor. When any annuciated function deviates from a specified condition, an auditory alarm is initiated causing one or more tiles to blink. Separate buttons cancel auditory and blinking signals. When the blinking signal is canceled, the tile stays illuminated in steady-on condition until the trouble is solved. Generally, 20 or more tiles are in steady-on condition during normal operation, due to various events that do not require immediate action. At the beginning of the accident, numerous alarms are initiated (more than 100 were immediately initiated during the TMI accident, but only few of them were essential for the event diagnosis); thus, they contribute to stress and confuse the operator if they are not properly designed and ordered. With the fast development of new electronic and computer-based technologies, new concepts and equipment have been introduced in control rooms as operational support systems. Cathode-ray tubes (CRTs) are now commonly used to supervise and alarm the plant. CRTs are the visual interphase to the operator, e.g., in the case of the Safety Parameter Display System (SPDS) used in most modern nuclear plants (it is a display of selected safety-related parameters). Its primary function is to help the operator to quickly assess the plant safety status by controlling the main safety functions. According to what the operator selects, it may offer rough data or a synthesis of the plant status on one or more CRTs. Information is presented in many formats, which include bar graphs, polarographic and mimic displays, and time histories. Most of operating nuclear power plants have backfitted SPDS in their existing control rooms as a very efficient means to improve the operator's performance, especially in the diagnosis phase.

After making basic design decisions on the type of instruments, display and control instrument specifications are prepared. The human factor efforts help to ensure the provision of displays and controls capable of supporting the task performance required. Effective display and control designs facilitate a reliable information transfer between the personnel and the sensor or the actuating elements designed in equipment components. Display/control movements shall be compatible with personnel expectancies.

Scales represented on continuous displays or controls shall be easily translated, while an appropriate coding enables the rapid recognition of the acceptability of displayed values. In addition, displays and controls incorporate other features that shall promote the effective control of plant operations and reduce maintenance task demands.

Finally, a communication system must provide a means to monitor and control the plant operations not represented on control room displays and controls as well as meet other needs in terms of information exchange. Human factor efforts in the communication system design produce recommendations aiming at locating communication nodes throughout the plant and providing an adequate number of channels to exchange information. In nuclear power plants, communication systems usually include paging and conventional and sound-powered telephones and radios. An effective system integrates the system equipment capabilities and permits the message sender to select the receiver, alert the receiver of an incoming message, and have a sufficient number of channels to exchange intelligible information.

VI. DESIGN FOR MAINTAINABILITY

Experience shows that to improve the plant availability, safety and productivity it is necessary to improve the maintainability design of plant facilities, systems, and equipment as well. Long outages, incidents, and personnel injuries are often caused by a poor maintenance, due to the lack of attention to human factor principles in the design phase. A general consensus is largely shared on the need of an integrated approach to human factors within the framework of a comprehensive program plan, derived from the same general guidelines defined in Section IV.

In this section we will present only a few considerations on the topic; for a more detailed analysis, see the References. In general, we can divide our comments into overall plant guidance, system and equipment and support provisions guidelines. The first item addresses the design of plant facilities, with special attention to maintenance workshops. The following factors should be considered.

Plant design factors — It is necessary to make a systematic analysis of design and evaluation processes to tailor the plant to maintenace needs. Among the various factors to consider are

1. Identification and definition of buildings and facilities to support plant maintenance
2. Provision of sufficient space and volume for maintenance purposes
3. Arrangement of buildings and facilities on a maintenance basis
4. Location of workshops consistent with the plant site characteristics and planning of maintenance practices
5. Planning of access routes for the transport of materials to and from the yard
6. Planning of yard and shop space for increased outage requirements

7. Development of specific transport requirements for all major equipment
8. Possibility of changing maintenance needs during the various phases of the plant construction and operation or because of the plant aging

Workshop design — Consideration must be given to the workshop size, clearances, access provision, lift and transport aids, storage, workbenches, laydown spaces, comfort and environment, office and administrative spaces, hot workshops, and their arrangement.

Hazard and protection aspects of facility design — Design measures must be included to prevent the most common industrial accidents related to maintenance activities.

Environmental factors — The most important environmental factors to take into account are temperature and humidity, noise and vibration, and illumination.

It has been suggested that such factors heavily influence error rates. When the effective temperature (measure of subjective warmth, when air movement, humidity, and temperature are considered) reaches 32°C, decision-making errors significantly increase. Burst of noise of 110 dB have been shown to increase decision-making errors until 20 s after they occur.

Movement of people and equipment — It is necessary to make an analysis as to the safe and efficient movement of people and equipment during normal operations and outages. Study items will include elevators, cranes, hoists, platforms, catwalks, scaffolding, manual and powered vehicular devices, stairs, ladders, and ramps.

Maintenance communications — The importance of communication systems is largely due to the difficulty in personnel movements, high cost of downtime, and safety and availability impact of a poor coordination with management and operating staff. The most common problems involve the plant compartmentalization, noise level and protective clothing.

The second group of comments relates to system and equipment guidelines, including the following.

Equipment and system maintainability — An easy maintenance depends on standardization, interchangeability, modularization, connectors, fasteners, and diagnostic and testing features.

Preventive and predictive maintenance — The frequency of setting and adjustments as well as the early detection of failures will help to reduce human interventions and orderly plan the activities.

A third final group of issues is related to support provision guidelines and includes the following.

Technical information — It is very important to have the proper documentation on the updated condition of systems and equipment. Therefore, designers and suppliers must provide such documents while a configuration control system shall incorporate all plant modifications, equipment replacement and safety criteria changes.

Labeling and coding — Many incidents derive from maintenance troubles caused by poor labeling and coding; an extensive use of psychological and physiological considerations should be made; for instance, highly contrasting surface colors should be used both for color normal and color-defective workers.

Tools, stores and test equipment — Provisions should exist as for the location, arrangement and administration of tools, spare parts, and test equipment.

VII. PERSONNEL SELECTION AND TRAINING

As mentioned previously, personal skill, ability, attitude, education, and training are important internal PSFs.

In order to optimize the final results, the following steps should be followed.

A. Personnel Selection

The personnel selection should be based on a clear evaluation of the tasks to be performed.

In turn, this factor should define the educational curriculum, personal attitudes, and psychological characteristics. Selection should be generally based on:

1.   Selective tests, to actually ascertain the professional qualification of the candidate
2.   Medical examinations, to ascertain the physical fitness of candidates
3.   Aptitude and psychological tests, to gain information on the aptitude as far as the character and personality of candidates are concerned. The kind, extent and execution of these tests are very different from case to case.

B. Personnel Training

Training should be both theoretical and practical. Theoretical courses should be organized in order to ensure a clear understanding of the processes that will be controlled. Basic hydraulics, thermodynamics, electrotechnics, and electronics are generally required. Chemistry is also required for specific applications. The practical training is partly made in the plant and on simulators, when available.

The simulator technology has advanced together with the increased capability of analogue, hybrid, and digital computers. Since the 1950s, some analogue simulators have been used in the nuclear industry to train operators. The first example of nuclear simulator, which had a control desk identical to that of the real station, is the Calder Hall simulator designed and constructed at the U.K. Atomic Energy Research establishment in 1957. During the 1970s, most nuclear vendors and utilities worldwide adopted training simulators.

The currently available simulators may be classified as follows:

1.   Basic principles or concept simulators, with the purpose of illustrating general concepts, demonstrating and displaying the fundamental physical processes of a plant. They are used as an aid to the theoretical learning and for the preliminary training.
2.   Full scope simulators, simulating the whole plant. However, not all the plant systems are represented with the same degree of accuracy. In addition, it must be realized that behind the operator desk there is a mathematical model instead of the real plant. Thanks to computer technological advancements, simulators are ever more complete.

As far as the operator training is concerned, we would also mention the engineering simulators (although they are rarely used); they represent the entire plant but are not required to respond in real time since they are not linked to a real control room desk. They can be used for the plant dynamics understanding and for the training of engineers.

In 1979, the TMI accident highlighted the limitation of the simulators used in that period in the simulation of complex and unexpected accidents. The same TMI accident could not be reproduced on the B&W (the nuclear system designer). Present simulators have much larger capabilities.

Different opinions exist on the need of an "exact" replica of a given plant control room, or on the adequacy of a "similar" control room, where the major emphasis is put on processes, required actions and the timely understanding of the accident. A very important problem for modern, automated industrial plants is the "dual" training that enables the operator to act both as supervisor when the system is automatically working, and as manual controller during emergencies or degraded conditions. These two roles are not necessarily compatible or complementary; each of them requires different skills and knowledge.

C. Personnel Retraining

The operator's ability to cope with an accident is delayed depending on the period of time elapsed since his training. Fortunately, while after the initial training the normal activities are learned through the daily experience, accident operations may be refreshed through periodical retraining, particularly on simulators. Nuclear plant operators follow refreshment courses (retraining) approximately 2 weeks/year. Both in training and retraining courses, an important aid is given by the review of the operator's experiences. Real case accidents or near misses provide a valuable input in terms of training and retraining program definitions and offer interesting practical examples during the training sessions.

D. Maintenance Personnel Training

On a general level, insufficient attention has been paid to the maintenance personnel training. The so-called latent human errors are predominantly caused by unrevealed errors during maintenance or testing.

Greater emphasis is expected on this matter in the future.

VIII. OPERATION AND EMERGENCY PROCEDURES

The TMI accident has demonstrated, at least for nuclear power plants, that the classical guidance provided to operators in order to mitigate the accident consequences may be inadequate when multiple incidents, failures, or errors occur during or after the accident. The previous event-oriented procedures entailed a correct diagnosis by the operator (choosing among events included in the procedures) and the following step-by-step plant recovery, based on the idea that the accident would not significantly deviate from the design hypothesis. New symptom-oriented procedures have been developed, which do not require a precise diagnosis of the event but tend to keep the plant safety parameters within an acceptable range.

More specifically, with reference to the "Emergency Response Guidelines" developed for PWRs Westinghouse, guidelines consist of two independent sets of procedures and a systematic tool to continuously evaluate the plant safety through the response to an accident. The first set comprises the "optimal recovery guidelines", which are entered every time the reactor is tripped or the emergency core cooling system is started. When the accident nature has been identified, the operator is transferred to the applicable recovery procedure or subprocedure. A permanent rediagnosis is made throughout the application of guidelines and cross-connections are provided to the adequate procedure whenever a diagnostic error is identified. The optimal recovery guidelines are "scenario oriented". Moreover, in the initial phase of the accident, the operating staff monitors the critical safety functions. These are defined as a set of functions ensuring the integrity of the physical barrier against radioactivity release. The function restoration guidelines are entered when the critical safety function monitoring identifies a challenge to one of the functions. Depending on the severity of the challenge, the transfer to a function restoration guideline can be immediate or delayed. Such guidelines are independent of the event scenario; they are only based on plant parameters and equipment availability. It has been demonstrated both analytically and through experiments on simulators that the new procedures help to drastically reduce the operator error probability in complex situations.

IX. METHODS FOR THE QUANTIFICATION OF HUMAN PERFORMANCE

There are several reasons why quantification is important. After establishing the range of human failure modes existing in a given situation, it is necessary to establish the likelihood of the different failure scenarios. This process helps the identification of weak areas in a system. Thus, if a design factor is revealed by the analysis as critical for the overall system safety, then a high cost penalty will be tolerated to maximize the potential reliability of the man-machine interaction.

Further, quantification is necessary to evaluate the overall system reliability in relation to some external risk criteria. In order to evaluate the man-machine system, it is necessary to ensure compatibility between the measures used to represent human reliability and hardware. Hardware reliability data are usually expressed in probabilistic terms. These can take the form of density functions derived from the mean times between failures, or point estimates of the overall probability of failure obtained from failure frequencies over a given time interval. This latter approach is commonly used to estimate the human failure probability, since it is very difficult to obtain time-based data on human error.

Three broad categories of approach to the human reliability quantification can be identified. Of these, the first primarily relies on the combination of historical data on the probability of failure for relatively basic elements of the human behavior (such as operating switches, closing valves, or reading dials) to obtain the probability of error for more complex tasks, which are aggregations of these basic elements. Such techniques are variously referred to as "analysis and synthesis", "reductionist", or "decomposition" approaches. The second category tries to apply the classical reliability techniques of time-dependent modeling to predict parameters such as the mean time between human failures. The third category makes greater use of quantified subjective judgment to supplement the currently inadequate data base of objective data on the probability of human error for various types of task. Also, these methods tend to take a more holistic approach to the evaluation of a task than the decomposition techniques.

We will limit our discussion to one quantification methodology, belonging to the "decomposition" category. It is essentially based on the assumption that human behavior in an NPP setting is procedure based (i.e., "by the book") as in the military case. However, most actions made by human beings to operate and maintain a nuclear power plant can be described as procedural. The procedure may be externalized (i.e., a written step-by-step list) or internalized (i.e., based on acquired skills).*

On the other hand, decision-making or problem-solving errors (less frequent but usually entailing greater consequences) cannot be modeled and quantified in a completely satisfactory manner at the present state of the art.

On a conceptual level, the general "decomposition" method for the analysis and improvement of human performance consists of the following steps:

1. Identification of all the interactions of people with systems and components, i.e., the man-machine interfaces
2. Analysis of these interfaces to see if PFSs can adequately support the tasks to be performed
3. Identification of potential problem areas in equipment design, written procedures, plant policies and practice, people skill, and other factors probably resulting in human errors
4. Decision on which problem could have a potential impact on the system to ensure changes
5. Development of candidate solutions to the problems
6. Evaluation of the estimated consequences of such changes to ensure the improvement of the system reliability and availability of safety systems and to guarantee that no additional serious problem will arise from them

This general method, which has been used for some time, is called man-machine system analysis (MMSA). MMSA can be used both as a qualitative and quantitative analysis. The qualitative part is based on a descriptive and analytical technique called task analysis. The quantitative part uses a human reliability technique to estimate the effect of human performance on system criteria such as reliability and safety.

---

* The experience at Chernobyl stresses the importance of written procedures, assigned personal responsibilities, hierarchical organization of the staff, and penal laws at different levels.

For both qualitative and quantitive estimates, tasks performed by the plant staff are analyzed to identify actual or potential sources of human errors, i.e., error likely situations (ELS) or accident-prone situations (APS). Some techniques to identify ELSs and APSs in a complex man-machine system were developed by Dr. R. B. Miller and associates at the American Institute for Research in early 1950s. They have been refined and expanded to be applied to HRA. In particular, see the task analysis in Reference 7.

The technique for the human reliability analysis we will describe is called Technique for Human Error Rate Prediction (THERP). This section only discusses single point estimates of basic and conditional human error probabilities (HEP).

At the Sixth Annual Meeting of the Human Factors Society held in November 1962, the acronym THERP was first used to designate the human reliability method developed at Sandia National Laboratory (SNL). It can be defined as a method to predict human error probabilities and evaluate the degradation of a man-machine system likely to be caused by human errors alone or in connection with equipment functioning, operational procedures, and practices, or other system and human characteristics that influence the system behavior.

The method uses the conventional reliability technology with some modifications relating to the greater variability, unpredictability, and interdependence of human performance as compared with equipment. THERP steps are similar to the conventional reliability analyses, except that human activities replace the equipment output. The steps are as follows:

1. Definition of the system failures of interest. They relate to system functions that may be influenced by human errors, for which error probability is to be estimated.
2. Listing and analysis of the related human operations. This step is the task analysis described in Section IV.
3. Estimate of the relevant error probability.
4. Estimate of the effects of human error on the system failure events. This step usually involves the HRA integration with a system reliability analysis.
5. Recommendation of changes to the system and new calculation of the system failure probability (the procedure is iterative).

The above five steps make use of HRA as a tool in the system design. For assessments only, step 5 is not required; not all tasks will be analyzed and included in the HRA. For PRA purposes, HRA will include only those tasks that have a material effect on the system event or fault trees.

The basic tool of THERP is a form of event tree called probability tree diagram, in use at the SNL since the 1950s. In the HRA event tree, the limbs represent the binary decision process, that is, correct or incorrect performance are the only possible choice.

According to the above-described general guidelines, the following modeling could be feasible.

Each operator action is modeled to consist of three phases:

1. Diagnosis (and decision to act)
2. Action
3. Recovery (when the action phase fails)

No recovery of the diagnosis failure is modeled.

For each of these phases, a failure probability can be calculated using THERP methods and other human error event trees, if necessary. If one labels the failure probability of each phase by:

Q_D = failure probability of the diagnosis phase

Q_A = failure probability of the action phase

Q_R = failure probability of the recovery phase

then the HEP failure probability of the operator action can be calculated as

$$\text{HEP} = Q_D + (1 - Q_D)Q_A Q_R \tag{1}$$

In most cases, $1 - Q_D = 1$, and the above equation becomes:

$$\text{HEP} = Q_D + Q_A Q_R \tag{2}$$

In HRA, a major problem consists of determining how the probability of failure or success of one task may be related to the failure or success of some other task. Two events are independent if the conditional probability of one of them is the same, whether the other event has occurred or not. Failure to make a realistic assessment of dependence may lead to overoptimistic evaluations on the joint human error probabilities. Dependence may occur between and within people. Dependence between people is fairly common, e.g., one person restores some manual valves while another checks the accuracy of the restoration. Generally, some dependence will arise in this case. Dependence also occurs between different tasks performed by the same person. For instance, dependence is likely to occur when a person restores two adjacent valves or activates two different adjacent switches.

The paucity of actual data on human performance continues to be a major problem to estimate HEPs and performance times. Therefore, the analysis should turn to various sources of information to derive such estimates for HRA purposes.

To estimate HEPs, the usual sources are extrapolation combined with expert judgment or the use of expert judgment only. The estimation of performance times poses a more difficult problem, since task times taken from different types of plant (i.e., nonnuclear data used for nuclear plants) cannot be used for accurate predictions. The best approach consists of measuring actual or simulated task times. The following source categories are used:

1. Nuclear power plants.
2. Dynamic simulators.
3. Process industries.
4. Job situations in other industries and in the military field, psychologically similar to Nuclear Power Plant tasks.
5. Experiments and field studies using real-world tasks of interest, e.g. experimental comparisons of measurement techniques in industrial setting performance records of industrial workers, etc.
6. Experiments using artificial tasks, e.g., typical university psychological studies, which have a limited application to real-world tasks.

The above list orders the sources according to their relevance to Nuclear Power Plant tasks. Unfortunately, the availability of HEP data is almost exactly the opposite.

In addition to HEP estimates based on such "hard" data, predictions are prepared according to expert judgement. This method is used in the error probability modification to judge the extent of known probabilities when they have to be applied to some different situations.

REFERENCES

1. An analysis of root causes of significant events identified in 1985, INPO 86-022, September 1986.
2. Reactor Safety Study, Wash 1400 (NUREG-75/014), October 1975.
3. Handbook of Human Reliability Analysis with Emphasis on Nuclear Power Plant Applications, NUREG/CR-1278, August 1983.
4. Human Engineering Design Guidelines for Maintenability, EPRI NP-4350, December 1985.
5. Human Factors Guide for Nuclear Power Plant Control Room Development, EPRI NP-3659, August 1984.
6. PRA Procedure Guide, NUREG/CR-2300, January 1983.
7. A Procedure for Conducting a Human Reliability Analysis for Nuclear Power Plants, NUREG/CR-2254, 1983.

Section II
Conventional Risks

INTRODUCTION

High-risk plants or, however, plants handling high-risk substances may be the origin, par excellence, of high risks. To high-risks limitation Chapters 1 to 5 have been aimed.

The design of an industrial plant, even a high-risk one, nevertheless must take into account what may be called "conventional risks", which, generally, only regard personnel involved in operation and maintenance and not population, but which are by far the greatest origin of occupational diseases and deaths.

Such conventional risks are here mentioned both because of their relevance from a general point of view, and because of the relevance that human factors have shown in the reliability of high-risk plants (see Chapter 6), that suggests to take into the maximum account the physical and psychological health of personnel.

This may be achieved also limiting the causes of conventional risks.

In some cases conventional risks may represent the starting events of accidental chains which endanger the overall high-risk plant.

Chapter 7

CONVENTIONAL RISKS

Rinaldo Paciucci

TABLE OF CONTENTS

I. INDUSTRIAL CONSTRUCTIONS

Industrial constructions must be built to create the best possible working conditions with the fewest possible physical and chemical risks. The following features must be considered in these buildings:

1.   Illumination, microclimate, environmental noise, and limitation of vibrations
2.   Overall structure: sound enough to resist the effects of production techniques employed
3.   Space: sufficient for the smooth operation of machinery and installations
4.   Accesses and exits sizable enough to allow for smooth, direct, and safe transit of personnel, production materials, and goods

Industrial constructions can be of one or more stories though one story constructions offer definite advantages: far easier access, floors that can hold up under heavy loads, and simple rapid exits for personnel. However, they also occupy a vaster area and there is greater heat loss through dispersion.

Buildings of more than one story allow for better utilization of territory and less heat loss.

For efficiency and administrative reasons, they are a positive solution for light industries, office buildings, and laboratories where no special risk factors are present. It is usually better to carry out all phases of production in the same building.

Factor such as hazardous or highly polluting production techniques linked to specific phases, or local zoning laws, may make the construction of more than one building necessary.

A. Accesses and Exits

Accesses must guarantee smooth, direct, safe transit to personnel and vehicles. They must be as linear as possible and wide enough to handle the necessary amount of traffic. Accesses for personnel only must be at least 1.20-m wide; 60 cm must be added for every 50 persons after the first hundred. These accesses must be free of obstacles of any kind.

Access for vehicles and personnel must be wide enough for both, 70 to 80 cm wider than the widest vehicle. They must also be clearly marked on the pavement. Exits, i.e., passages from the place of work to "safe" places (out-of-doors or sheltered areas), should never be more than 30-m long, less if production is hazardous. Proper planning should include two exits for every place of work. The exits should be marked with signs and phosphorescent arrows. Accesses and exits should be protected from falling weights and shrapnel.

All emergency controls (such as switch boxes) and first aid stations should be placed along the exits so that dangerous machinery and installations can be switched off easily even during rapid evacuation of the premises.

Wrong solution   Incorrect solution   Right solution

FIGURE 1. Stairs.

Easy to see and read maps of the building should be placed near the exits showing:

- The observer's location
- Location of fire extinguishers, switch boxes and all other emergency controls
- The closest and fastest exit

All doors along exit passages must open out.

B. Stairs

In buildings of more than one story there must be stairs and elevators. Stairs can also be seen as emergency exits. Their position and width depends on the same factors we considered for entrances except where there is greater need for evacuating large number of people. Then wider staircases are necessary.

A 45° slope with no more than 12 stairs before each landing is recommended for linear ramps (see Figure 1). Step widths are determined on the basis of the following formula (data in cm):

$$63 < s + 2 h < 65 \qquad (1)$$

where s = step and h = height.

Staircases should run along the outside walls of buildings but this is obligatory only in case of severe fire hazards. Whenever fire is a serious and real risk factor staircases should be built in towers a few feet from the main building linked to it by passageways. "Smokeproof" staircases completely surrounded by a 40-cm thick wall and accessed only from the roof or from well-aired balconies with fireproof, smokeproof automatic doors are another alternative.

Smokeproof staircases must be ventilated through an opening more than 1 m^2, situated above the roof entrance.

Staircases must have banisters, lead directly out-of-doors, and be closed off from elevator hoists, garages, and basements used for storing flammable goods.

II. PROBLEMS WITH PRESSURIZED GAS

Areas of production with high risk of explosion, poisoning, and fire must be closed off by special resistant partitions for the full period of the reaction time of the critical agent. A special installation outside the main plant, if possible, should be built with well-ventilated walls and interstices crossed by steel pipes containing gas pipes, exhaust ventilators in the most hazardous areas, fireproof electrical wiring, adequate safety measures in case of explosion such as roof, and wall openings normally covered with panels which open easily when there is a rise in pressure as in an explosion.

When using pressurized gas it is better to keep the tanks in a central, specially constructed distribution area, preferably outside the main building.

All areas crossed by gas lines must have special detectors geared for the specific gases used with a double safety gauge: alarm which goes off at one third of explosion level and automatic shut-off at one half.

Gas lines in stainless steel painted in specified colors should be in full view (especially shunts) and slightly distanced from the wall to limit corrosion in occult areas.

Manually operated shutoff gauges and valves are to be installed where gas lines begin and end.

The gas distribtion center should be made up of individual compartments built of non-flammable, nonconducting materials (such as concrete). Roofing must offer the least possible resistance to explosion (eternit).

Each totally isolated compartment should contain only one tank firmly clasped or chained to the wall. Empty tanks must be kept separate from full ones. Incompatible gases must also be kept carefully separate (flammables and combustion supporters).

Concrete trenches must be built for gas lines crossing open ground (one for each group of compatible gases). The lines are then attached to the side walls of the trenches using the precautions described above. The trenches, covered to reconstruct original surface, must be well drained.

III. CRITERIA FOR PLANNING AREAS WHERE HIGH-RISK FACTOR SUBSTANCES ARE TO BE USED

So as to limit the probability of extending risks to areas where no special precautions have been taken, production phases and/or machinery with high risk of explosions, fires, or production of poisonous gases must be isolated within specially protected areas or specially constructed buildings at a safe distance from other constructions.

In laboratories and installations where highly toxic substances known for causing severe accidents (like phosphina TLV = 0.3 ppm, silano, phosgene, etc.) are used, a series of measures to cut risks even when minimal amounts are present must be included in the planning phase.

A special area for the use of these products must be found and built so as to permit fireman and emergency teams immediate action.

The area chosen should have a container (like a glove box) for the production unit and tanks of the toxic substance, allowing enough room for smooth operating procedures. The pressure in the container should be slightly lower than normal (a few millimeters of a water column) and have fan motors fed by an autonomous power supply in case of blackout.

The fan vent must have a charcoal filter so, in case of accident, this filter absorbs the poisonous substances usually deposited in a vat.

Under normal conditions, exhaust can be let out through a bypass to keep the charcoal from getting damp.

Switching to the filter vent must be automatic (manual control should also be provided) whenever the detectors in the box measure the given concentration of poisonous gas. Switching signals will trigger cutoff valves (pneumatic or antideflagration for flammables), set along the gas distribution lines near the tanks and the containers (boxes).

In order to keep poisonous gas/or flammable substances damages to a minimum, laboratories, plants, and deposit for these products must be situated in special areas so that none of the hypothetical effects be brought happen on nearby populated areas.

When working with flammable substances, the possible risks are scattered liquids catching fire or reservoirs of liquids catching fire and explosions occurring emitting a cloud of poisonous gases.

Table 1
EXPOSURE LIMITS FOR PERSONS
AND MACHINERY

Exposure time for persons	Thermal flow	
	BTU/h × ft²	KW/m²
Few seconds	3000	9.46
20 s	2000	6.30
2 min	1500	4.73
10 min	1000	3.15
2 h	500	1.58

Note: Machinery must be set up so as not exceed 38 kW/m² in stock containers and 12.5 kW/m² in wooden or plastic structures or areas open to the public.

The consequences of exposure for men and objects can be seen in Table 1 which gives maximum exposure times for a normally dressed person as a function of a given heat flow. All apparatus must be set up keeping values within the following limits:

1. 38 kW/m² (12000 BTU/h ft²) in stock containers
2. 12.5 kW/m² (4000 BTU/h ft²) in wooden or plastic structures and in areas open to the public

The consequences of explosions can be seen from Table 2 which shows the hardest hit areas as a function of distance from the epicenter and extension of the cloud of poisonous gas (in equivalent tone of hydrocarbons).

Efficient emergency programs must be set up in agreement with local authorities. Emergency plans do not replace safety measures but are an additional guarantee for the surrounding areas as well as for workers themselves in case of accidents.

IV. LIGHTING

A. General Comments

Lighting is one of the most important environmental factors having a direct effect on personnel. When it is inadequate, it creates unsatisfactory physiological conditions and thus physical and sensory fatigue. Proper lighting, whether natural or artificial, is beneficial to everyone and creates the essential environment for carrying out the work at hand.

Having enough high quality well-distributed light is not just a whim but a factor that makes working easier and boosts production.

Proper lighting as a boost to production was the factor that triggered research in the field of lighting science. The first experiments, carried out on a group of printers in England in 1926, show that an increase in lighting to just above 200 lux leads to an increase in production of up to 20%. These results were confirmed by experiments carried out in the U.S. on groups of printers (Figure 2).

With the birth of fluorescent light (much stronger than incandescent light) experimentation spread to work situations characterized by very strong lighting. A study carried out in Germany (Figure 3) shows that as light intensity increases so does production. A saturation point is reached at about 2000 lux.

Since results were evaluated not only on the basis of work done but on number of errors made and degree of eye fatigue, lighting was proven to be an important factor in job safety and accident prevention.

Table 2
AREAS NEEDING SPECIAL PLANNING PROCEDURES

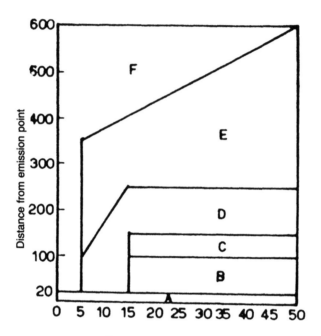

Extension of uncontrolled clouds
(ion equivalent hydrocarbons)

Area	Procedure
A	No inhabited buildings
B	No roads suitable for vehicles
	No high-risk installation (domino effect)
	Buildings planned for peak pressure of 10 psi for 20 ms
C	Buildings planned for peak pressure between 3 and 10 psi
D	Buildings planned for peak pressure between 1.5 and 3 psi
	No roads
E	No homes
F	No limits on the project

General fatigue is greatest when lighting intensity is less than 50 lux. General fatigue decreases as intensity increases up to 1000 lux then increases as intensity increases.

Today lighting intensity is chosen on the basis of angle of vision respect to the object observed (its size and distance from the observer, Figure 4), perception speed (see Figure 5), and cost of electricity.

European standards are lower than American standards though ensuring the same visual acuity. Table 3 shows lighting intensities advised under the following conditions:

- Immobile objects
- Background brilliancy 0.8 (about equal to that of this paper)
- Natural attention without efforts at concentration
- White fluorescent light of 3500 K.

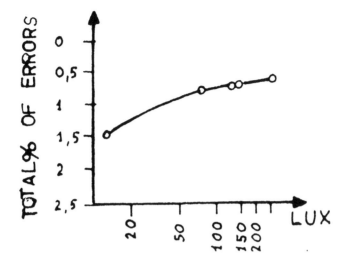

FIGURE 2.   The number of mistakes as a function of light power.

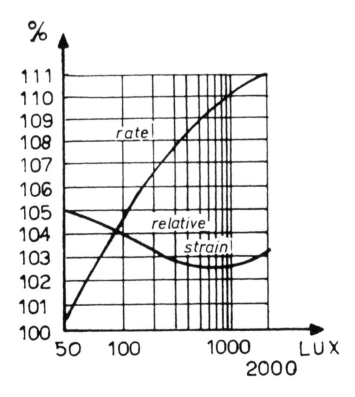

FIGURE 3.   Efficiency as a function of light power.

Correction factors for different situations are found in Table 4. Knowing about the best lighting intensities is not enough to ensure "good" lighting. In fact, technological advances over the last 10 years have led to vast changes in job techniques, decreasing the need for physical force and increasing intellectual and physical activity. Therefore, visual activity has also changed requiring greater precision and leading to the identification of the criteria listed here for determining lighting.

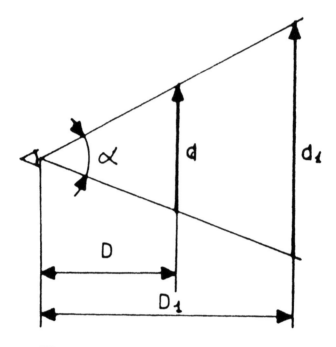

FIGURE 4.   Visual angle subtended by the object observed.

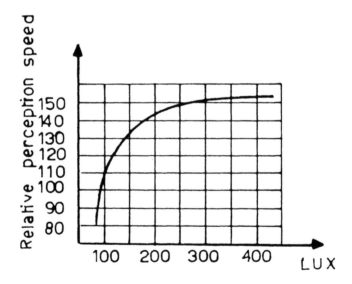

FIGURE 5.   Lighting power as a function of perception speed.

Field contrast — In the long run, frequent eye shifts from one object to another at different lighting intensities leads to eye fatigue because of the continuous need for the muscles of the pupil to adapt to the different situations. National standards require a 3:1 ratio in lighting between adjacent work areas and a 10:1 ratio respect to surrounding areas.

Discontinuity factor — This is similar to field contrast and is linked to overall lighting intensities where a no less than 2:1 ratio must be kept between lightest and darkest areas.

Shadow factor — This is the ratio:

Table 3
CHOICE OF LIGHTING STRENGTH AS A FUNCTION OF SIZE OF OBJECTS, CONTRAST, AND VISUAL ACUITY REQUIRED

Details to note	Apparent size		Applications	Relative visual acuity	Illumination (lux) CONTRAST			
	α	D/d			0.8	0.4	0.2	0.1
Very fine	0' 50"	4100	Watchworks, high precision work, surgery, visual limits	0.98	4300	7100	11000	20000
				0.95	2300	4850	8500	15500
	1' 05"	3200	Technical design, high precision work, weaving, printing	0.90	1350	3250	6100	13800
				0.80	405	1700	3500	12400
Fine	1' 25"	2400	Window dressing, exhibits, bookkeeping, schools, libraries, turning, milling, reaming	0.98	1810	4400	7450	11200
				0.95	850	2700	5100	9100
	1' 50"	1900	Offices, machine assembly, chemical food, paper industries	0.90	420	1700	3700	6400
				0.80	110	690	1350	3700
Medium	2' 20"	1500	Offices where work is irregular, archives, waiting rooms, chemical installations	0.98	295	1800	4000	6500
				0.95	125	880	2550	4600
	3'	1100	Presses, forges, lamination work, carpentry	0.90	38	360	1450	2800
				0.80	13.5	145	620	1450
Coarse	4' 30"	850	Ovens, hallways	0.98	64	540	1800	3300
				0.95	22	190	800	1900
	6'	600	Furnaces, cement mixers, coarse work	0.90	6	67	295	1000
				0.80	3	29.5	135	480

Note: Relative visual acuity is the relationship between acuity obtained and optimal acuity.

Table 4
CORRECTION FACTOR

Background luminosity	0.8	0.6	0.5	0.4	0.3	0.2	0.1	0.05
	1	1.3	1.6	2.5	3.5	5	10	20

Perception speed	Still	Slow	Fast	Very fast	Extremely rapid
	1	2	3	4	5
Attention	Natural	Normal	Intense	Very Intense	Spasmodic
	1	2	3	4	5—6
Daylight light color	Na vapors	Incandescence	Fluorescence	Hg vapors	Fluorescence
	0.4	0.8	1	1.5	2.5

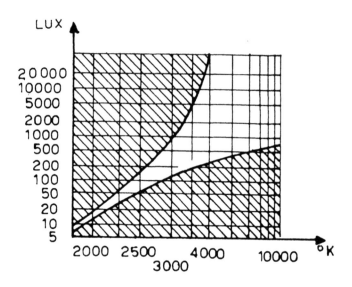

FIGURE 6.   Kruithoff's diagram.

$$K = \frac{E_2 - E_1}{E_2} \qquad (2)$$

where E_2 is the lighting intensity expressed in lux in directly lighted areas and E_1 is the lighting intensity value in areas where shadows are cast.

The shadow (shade) factor, though bettering vision of embossed materials, must be between 0.2 and 0.8 to avoid problems linked to field contrast.

Color temperature — This factor emerges when dealing with artificial lighting. It has been found that although lighting is adequate it may appear pleasant or unpleasant according its color temperature. The human eye is used to associating soft light with cool colors (dawn and sunset) and strong light with warm colors. If this harmony is broken, light is seen as unpleasant. Kruithoff's diagram gives a fair picture of the relationship between color temperature and lighting intensity in lux (Figure 6).

Index of color quality — The index of color quality (ICQ) expressed in percentages is the quality of color under artificial light as compared to the same in sunlight. It is an important factor in visual fatigue because if colors are untrue it takes greater care to identify them.

Stroboscopic effect — Appears when objects move rapidly under fluorescent light. It is extremely annoying and can also lead to false perception of movement. It is caused by the intermittance of flux of light at network frequency.

Luminance contrast — This is the relationship between light emitted by adjacent surfaces, for example, light emitted by a source and that reflected by the ceiling, or light coming through a window and that reflected by objects in the same visual field.

Luminance contrast should be as low as possible since the human eye is used to working with wide areas of low luminance contrasting but little with adjacent areas.

Artificial light sources are small areas with extremely strong luminance surrounded by zones in half light (ceilings). The use of fluorescent light has bettered the situation and the use of unshaded light sources creates vast high luminance areas around the source.

B. Natural Lighting

Since natural lighting varies in intensity with the time of day, the season, the latitude, and especially with atmospheric conditions, it is sensible to evaluate it in absolute terms.

Daylight coefficients are used to express the relationship between lighting on the work table (E) and the light outdoors (Eo) excluding the effects of direct sunlight.

The table of values allowed for η are found by assuming the values listed in Tables 3 and 4 for E and assuming Eo = 3000 lux, the value of natural light when the sky is completely clouded over.

Field contrast, discontinuity, the shadow factor, and luminance contrast must also be taken into account when considering natural light. Direct sunlight on the work table must be avoided to eliminate excessive field contrast and dazzling. Natural lighting E can be calculated using the architectural characteristics of the building. Equation 3 shows the overall flux of light in the environment (Φ):

$$\Phi_u = \frac{f \cdot \gamma \cdot A_{tot}}{1 - \delta_m} E_o \qquad \text{(lumen)} \tag{3}$$

Mean lighting is obtained from the overall flux and the floor area (Fa):

$$E = \frac{\Phi_u}{f_a} \frac{f \cdot \gamma \cdot A_{tot}}{(1 - \delta_m) \cdot F_a} \qquad \text{(lux)} \tag{4}$$

so

$$\eta = \frac{E}{E_o} \frac{f \cdot \gamma \cdot A_{tot}}{(1 - \delta_m) f_a} \tag{5}$$

A_{tot}	= total glassed in area,
$A_{wi}, \ldots A_{wn}$	= area of window 1, . . . nth,
δ_m	= coefficient of mean environmental diffusion (depends on the color and structure of internal surfaces; values have been tabulated,
fl, . . . fn	= geometric factors of window,
f	= factor of mean window, and
γ	= ratio between luminosity with and without glass.

The geometric window factor is calculated as in Figure 7.

$$f = \text{sen}^2 \frac{\alpha}{2} \tag{6}$$

where α is the angle delimiting the view of the open sky.

Equation 5 can be used to evaluate the glassed in areas needed during the design phase.

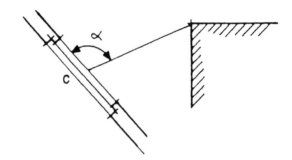

FIGURE 7.   Geometric window factor.

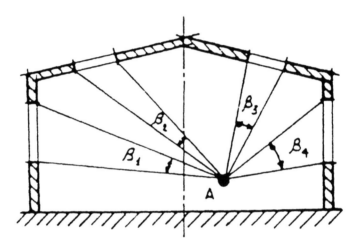

FIGURE 8.   Transverse section of the building with the indication of visible sky.

As for daylight distribution (coefficient of discontinuity), this can be tested using Buff's diagrams of percent of sky visible.

For each point on the working surface (1 m from the floor) the ordinates on the diagram are proportional to the sum of angles B_i formed by the different windows and point A. As a first approximation, the amount of sky visible (As) can be calculated using (see Figure 8):

$$A_s = \frac{\sum_i \beta_i}{\pi} \frac{L_w}{L_f} \tag{7}$$

where Lw = length of windows perpendicular to the plane and Lf = floor length in the same direction.

A diagram like that in Figure 9 can be made showing degree of illumination at any given point of the working area in a transverse section.

The relationship between As max and As min is dicontinuous. Skylights obviously render light distribution more continuous. To keep direct sunlight from shining on the work table producing negative field contrast effect, skylight glass should diffuse light.

Shed roofing on industrial constructions also eliminates the problems of direct sunlight. This roofing is made of two asymmetric sloping flaps, one of glass.

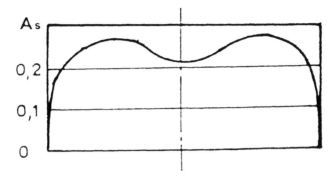

FIGURE 9. Diagram of degree of illumination at any given point of the working area in a transverse section.

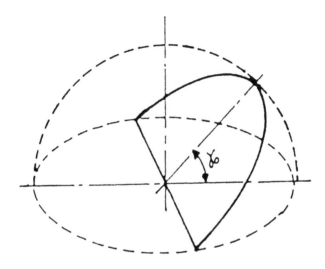

FIGURE 10. Inclination of elliptical plane on equatorial plane.

Sheds are to be built with the flaps of glass facing north cast on a slop (γ) of no less than:

$$\gamma = 90 - \phi + \epsilon \qquad (8)$$

where ϕ is the latitude of the site and ϵ is the telt of the equatorial plane respect to the elliptical plane (see Figures 10 and 11).

C. Emergency Artificial Lighting

Work entrances and staircases must have emergency lighting systems run by an independent power supply to ensure safe easy evacuation of premises in case of blackout. Emergency lighting must ensure 5 to 10 lux on the floor.

D. Lighting of Parking Lots and Loading Zones

Parking lots, loading zones, and other outdoor areas must be lighted to at least 10 to 15 lux.

E. Problems with Video Terminals

Aside from project design rooms, offices in general do not require visual precision.

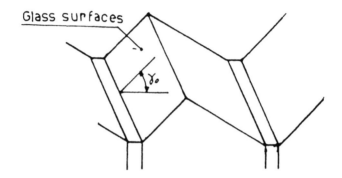

FIGURE 11.   Example of orientation of shed glass surfaces.

Many of the objects to be observed are two dimensional and do not need shading to be seen better. However, whereas it takes 300 to 400 lux to read papers, lighting for computer monitors must be much softer.

Aside from the adaptation difficulties of light reflected from keys, some kinds of paper can cause veil reflections contrast inversion and at worst annoying dazzling effects. Tolerable levels of reflected light are obtained increasing the diffusing surface (rough surface), using light filters and antireflection films.

It is best to choose the right lighting system and (arrangement) placement of personnel and objects respect to light sources.

Under these conditions diffuse lighting offers the following advantages:

- Decrease in visible light points
- Decrease in reflected light in visual field
- Decrease in contrast

A more continuous distribution of light intensity therefore results.

On the other hand, low efficiency raises production and administrative costs. It is therefore worthwhile to analyze the factors listed above, since they are fundamental to the work of a video operator.

Lettering contrast — This is the relationship between lettering luminosity L and that of the screen Ls. According to the law DIN number 66234 this should range between 6:1 and 10:1.

Screen contrast variation — This is measured as the ratio between nine areas of continuous luminosity generated on the screen and the surrounding areas. The limits are those given for lettering contrast.

Percentage reduction of lettering contrast —

$$R = 100 \frac{C_{max} - C}{C_{max}} \qquad (9)$$

This is the percentage ratio between contrast C_{max} obtained from a control lighting system (sphere or pinpoint source) and contrast C in the area studied against a contrast sample made of one black and one white element defined as:

$$C = \frac{L_{black} - L_{white}}{L_{black}} \qquad (10)$$

where L is the luminosity of samples studied from the observers view point. It is important to remember that contrast decreases are mostly due to veil reflexions. Contrast decrease on papers should not be more than 15 and 10% on the keyboard.

V. MICROCLIMATE AND AIR CONDITIONING

On the job microclimate should be adapted to the type of work done.

Microclimate is influenced by a series of objective climatic factors plus those factors which act on temperature change such as number of people present and degree and type of activity.

The human body has a heat regulation system which keeps internal temperature at around 37°C. To keep this temperature constant, produced or acquired by the body must be equal to the amount dispersed.

Heat produced by the body is metabolic heat, energy burned, and is the sum of two factors: basal metabolism which yields the heat necessary for vegetative life and heat produced by work. Heat given off by or taken on by the body comes from individual environment heat exchanges through conduction (practically negligible), convection, radiation, and respiration. Heat is also given off through perspiration evaporation. Heat balance can therefore be represented by the following formula:

$$\pm Q = M \pm Xp \pm Xc \pm Xr \pm RES - Xe \qquad (11)$$

where Q = heat taken on or given off by the organism, M = metabolic heat, Xp = conduction heat, Xc = convection heat, Xr = radiant heat, Xe = evaporation heat, and RES = respiratory heat

Homothermia is where Q = 0. This is a situation of comfort. Where Q < 0 or Q > 0 there is a rise or fall in workers' body temperature which leads to feelings of strain and, in extreme cases, stress.

Since body temperature depends so clearly on microclimate, this must be controlled in a simple way using microclimatic indices to evaluate environmental conditions and correlate these to work activities tabulated according to energy burned.

A. Microclimatic Indices

The most widely used index of heat stress is the WBGT calculated as follows:

$$WBGT = 0.7 \, Wb + 0.2 \, Gt + 0.1 \, Db \qquad (12)$$

$$WBGT = 0.7 \, Wb + 0.3 \, Gt \qquad (13)$$

Wb is the temperature as indicated on a mercury thermometer with wet bulb (kept wet using capillary action by a cotton sock with one end in distilled water and exposed to natural breeze) and Gt is the temperature measured by a mercury thermometer with the bulb in the center of a 15-cm diameter hollow copper sphere painted flat black on the outside.

Db is the temperature on a mercury thermometer with a dry bulb.

Formula 12 is valid for outdoors where the thermometer is exposed to direct sunlight, Formula 13 is for indoors or outdoors in a well-shaded position.

According to the American Conference of Industrial Hygienists (ACGIH), thermoregulation limits in WBGT for light work (metabolic heat up to 200 kcal/h), moderate work (from 200 to 350 Kcal/h), heavy work (from 350 to 500 kcal/h) are listed in Table 5.

Workers can be exposed to temperatures higher than those indicated only if a medical checkup is carried out and they have proven a higher than usual resistance to working in overheated environments. Work must be interrupted when body temperature goes above 38°C.

Table 5
THERMOREGULATION LIMITS

	Work (°C of WBGT)		
Rhythm	**Light**	**Moderate**	**Heavy**
Continuous work	30.0	26.5	25.0
Work 75%, pauses 25%	30.6	28.0	25.9
Work 50%, pauses 50%	31.4	29.4	27.9
Work 25%, pauses 75%	32.2	31.1	30.0

B. Air Conditioning

For the reasons listed above for thermal well-being or because of special production techniques, either partial (simple ventilation) or complete (cooler plus humidifier or dehumidifier) air conditioning is necessary.

Ventilation — Quantity of ventilation can be measured against the maximum amount of pollution tolerated or against the need to lower temperatures created by work done. If in the first instance we have z = volume of pollutant produced in the environment in unit time, y = maximum amount of pollutant in the environment y_0 = amount of pollutant potentially present in the fresh air, and V = volume of air to be changed in unit time, then:

$$V \geq \frac{Z}{Y - Y_o} \tag{14}$$

If the need for ventilation comes from excess heat production then:

$$V = \frac{Q + q}{\gamma_a \cdot C_p \cdot (t_i - t_e)} \tag{15}$$

where Q = quantity of heat produced in unit time, q = quantity of heat exchange with outside through walls, ti and te = temperature inside and outside, γ_a = specific weight of air at room temperature, and Cp = specific heat of air at constant pressure (0.24 kcal/kg°C).

Complete air conditioning — For thorough information on complete air conditioning see specialized texts. Remember that it involves changing air, plus adding or subtracting heat and humidity. Installations usually have a filter, a heat exchange system to heat air in winter and cool it in summer, a humidifying chamber, a water filter, another heater, and ventilators to move air through the outlet.

Air drawn from the environment is partially recycled after purification to utilize the heat it contains. Pure air is drawn in through an upper vent to avoid dust or through a lower vent with filters or ionized water.

VI. ELECTRICAL INSTALLATIONS

Electrical installations are all apparatuses and materials necessary for the distribution and flux of energy to the machinery run by it.

This equipment must be perfectly efficient in terms of function, costs and safety. Safety is the most important since electricity is used in the home as well.

A. Danger of Electricity

1. Effects on the Human Body

Tetany — Tetany is muscle contraction induced by electrical energy. Whenever an

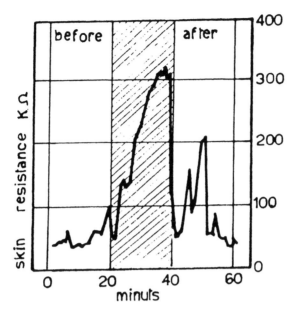

FIGURE 12. Human body resistance as a physiological condition of individual. The dark zone corresponds to intense concentration period.

electrical current of sufficient duration and intensity is applied to a nerve fiber, it propagates to the muscle ennervated which then contracts. The contraction ceases when the stimulus ceases. If stimuli follow each other in rapid enough succession the contraction will be continuous (tetanic contraction). When stimulus frequency is above certain values, the sum of the values results and the muscle goes into total contraction which lasts as long as the stimulus is applied. The muscle decontracts slowly to a resting state when stimulus ceases. A 50-Hz frequency is more than sufficient to induce the above state. If electrical contact continues, it may cause fainting, asphyxia, collapse, and death. This phenomenon occurs even at low voltages and is even more dangerous when one considers that prolonged contact with electrical current decreases the electrical resistance of the body. This decrease is also associated with the physiological and skin condition of the individual (Figures 12 and 13). When voltage is high enough, tetany does not occur because the stimulus is so violent that the injured person is sent flying backwards.

Ventricular fibrillation — Heart muscle contractions are normally regulated by impulses from the sinus auricolar node located in the upper part of the right atrium. When electrical impulses from an outside source interfere with these internal impulses, the heart muscle is stimulated chaotically and ventribular fibrillation, the cause of most deaths from electrical shock, results. Fibrillation is irreversible unless a defibrillator is applied.

Burns — Electricity passing through the body produces heat because of the Joule's effect due to the electrical resistance of the body. This is especially true at the contact points (where electricity enters and leaves the body). Severity of burns depends on duration of stimulus. Other alterations caused by electricity derive from its direct action on blood vessels, blood, and nerve cells. It can cause permanent heart, brain, or central nervous system damage.

2. Influence of Electrical Frequency

As a rule, the higher the frequency less dangerous is electricity. Animal cells are sensitive to variations in current.

Stimulus intensity necessary to excite the cell is inversely proportional to duration. Add

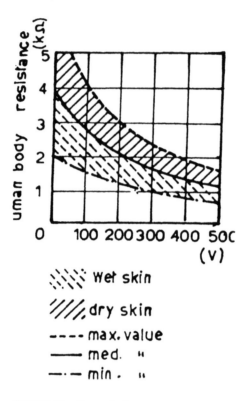

FIGURE 13.   Human body resistance as a function of tension (Carresscia, V. Impianti di Messa a Terra, ENPI).

the fact that high frequency current involves only the outers conducting areas (skin effect) running along the skin and leaving the vital organs intact. The thermal effects of high frequency current can, however, be very dangerous.

Figure 14 shows perception and tetany thresholds as a function of frequency on a 0.5 to 99.5% probability scale.

3. Electrical Shock Safety Curve

The curve of electrical values damaging the body as a function of time is the so-called safety curve. Unfortunately, statistical data applies only to individuals whose characteristics fall within ''mean'' values and therefore offers no absolute guarantee of safety or danger levels.

The IEC (International Electrotechnical Commission) indicates electrical danger points for individuals weighing 50 kg (this weight was chosen as a precautionary measure) whenever hand-hand or hand-foot contact is made (Figure 15).

The curve between zone 2 and zone 3 indicates the danger limit and derives from the equation:

$$I = 10 + 10/T \quad (mA) \tag{16}$$

where 10 mA (RMS value) is the limit before tetanization sets in (independent of time) and T is the time of contact in seconds.

For DC electrical current the following formula holds:

$$I_c = I_a * 1gT \tag{17}$$

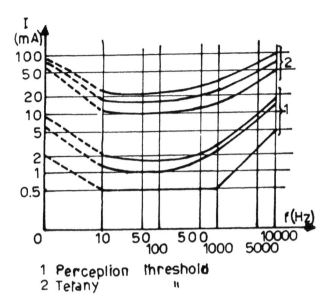

FIGURE 14.   Perception and tetany thresholds as a function of frequency on a 0.5 to 99.5% probability scale (Carrescia, V., Impianti di Messa a Terra, ENPI, 1974).

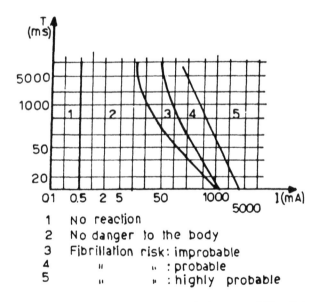

FIGURE 15.   Electrical danger points for individuals weighing 50 kg whenever hand-hand or hand-foot contact is made. By International Electrical Commission (IEC).

where I_a is AC value which produces the same effects as DC I_c and T is the time of contact in seconds. Figure 16 shows DC danger zones.

Danger limit for unique discharges (atmospheric discharges, condensers) is 50 w/s.

Once the danger limits for electricity on the human body have been found, we can find values for dangerous voltages by applying Ohm's law:

$$I = V_c/R_c$$

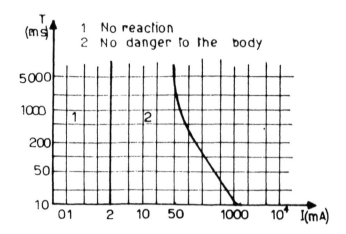

FIGURE 16.   Direct currents (DC) danger zones.

Table 6
VALUES FOR PROTECTION FROM
INDIRECT CONTACT AT LOW
VOLTAGES

Contact voltage (V)	Maximum contact (s)
< 50	Infinite
50	5
75	1
90	0.5
110	0.2
150	0.1
220	0.05
280	0.03

so, applying Formula 17:

$$V_c = R_c \times i = Rc(10 + 10/T) \tag{18}$$

where V_c = contact voltage, T is exposure time, and R_c is the resistance of the human body (see Figure 13).

The associated values in Table 6, taken from the curve in Figure 17, are those internationally (IEC) accepted for protection from indirect contact at low voltages.

Values quoted are lower for women, children, and men under 50 kg because of the lower electrical resistance of their bodies.

B. Grounding Systems

Grounding systems include conductors, joints, and dispersers to ground electrical leakage along a path of low resistance. Grounding systems ensure safe voltages for the full time of intervention of protective systems (fuses and switches) whenever there is a monophase breakdown sending current through a part of machinery where no current normally flows.

The safety grounding system includes dispersing elements touching the ground, ground conductors linked to the dispersing elements, and safety conductors to be linked to the metallic objects to be protected to the conducting elements on the ground.

Dispersing elements — Dispersing elements must be made of non corroding materials that guarantee efficient dispersion into the ground. Copper or copper- or zinc-plated iron may be used. Standard measurements for dispersing elements are given in Figure 18.

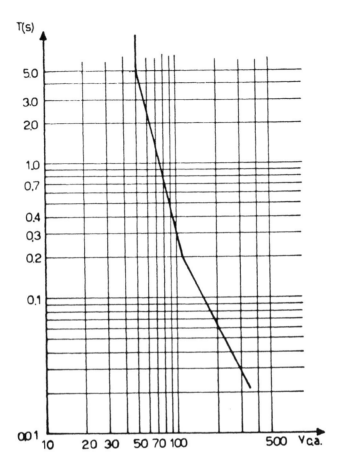

FIGURE 17. The maximum allowable duration of indirect contact as a function of voltage according to IEC.

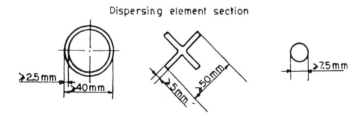

FIGURE 18. Dispersing elements dimensions according to Italian standard.

Picket-type dispersers are planted in the ground. Wire-like, spheric, or circular types are buried.

Grounding conductors and protection conductors — These are mostly made of copper but iron and aluminum are also used if the section is made thicker.

Elements for high voltage installations must be:

$$I/160 \quad \text{with at least 16 mm}^2 \quad \text{(copper)}$$

$$I/100 \quad \text{with at least 35 mm}^2 \quad \text{(aluminum)}$$

$$I/60 \quad \text{with at least 50 mm}^2 \quad \text{(iron)}$$

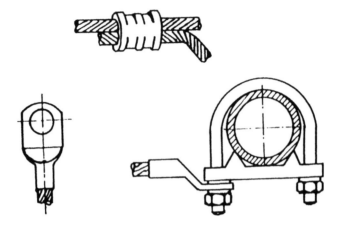

FIGURE 19.   Correct joints for grounding systems.

I is the maximum current the grounding system will ever need to disperse.

In low voltage installations, grounding and protective conductors must be equal in section to phase conductors up to a $16 - mm^2$ section.

When phase conductor sections are larger than 16 mm² grounding and protective conductor sections must be one half their values though at least equal to 16 mm². These measurements hold for copper. When the protective conductor is not in the same cable with the phase conductor it must be at least 5 mm² thick.

In case of doubt, calculate the section on the basis of heat tolerated, maximum electrical leakage possible, and time of protective element intervention if the phenomenon is adiabatic. Thus the formula:

$$s = I/\alpha = \sqrt{t/\Delta\theta} \tag{19}$$

where s = conductor section in mm², I = effectual value of electrical leakage in ampere, t = time of protective element intervention in seconds, α = coefficient typical for metals and equal to: 13 for copper, 8.5 for aluminum, 4.5 for iron, 2.5 for lead, and Δθ temperature reached by the conductor equal to 180°C for nude cables.

Joints — Joints must guarantee good mechanical and electrical flow efficiency. Contact surfaces must be large (200 mm² for the disperser-conductor joint at ground level). Joints must be either welded, hard brazed or closed with binding clamps (see Figure 19) to avoid fusion during the passage of current. Joints must be protected from corrosion by protecting them from humidity and avoiding contact between metals far from each other on the voltage scale (copper and zinc).

Geometry of grounding systems — Grounding systems limit voltage towards the ground through low ground resistance. Safe contact (hand to hand, hand to foot) and voltage of ground contact (foot to foot) voltages are maintained by creating equipotential connections and by placing dispersers correctly. The grounding system must cover the entire area to protect equally well. Wherever there is a great deal of electrical leakage, aside from the outer ring, there should be an inner network with a density corresponding to the need to keep contact voltages down.

The network should be roughly equal to:

$$Vco = 0.7 \; \rho I/L \tag{20}$$

where Vco = contact voltage, ρ = ground resistivity in ohm/m, I = maximum ground current in A, and L = conductor groundwork in m.

FIGURE 20. Symbol of double insulation.

For the outer edges of the network where voltage is higher:

$$Vpa = 4 \, \rho I/D^2 \tag{21}$$

where Vpa = voltage of step and D = diagonal of groundwork in meters.

The network is usually buried from 0.5 to 1 m deep because the dispersive capacity usually decreases with depth underground. However the dispersers must be safeguarded against outside effects (vehicle transit, ploughing, etc.).

C. Electrical Equipment

General instructions — Electrical equipment must meet legal standards, be suitable for the environment, and for energy distribution. The components of the installation and the machinery using electricity must be installed so as to ensure smooth running, control and use. Actioning and protective-control devices must be easy to recognize. Conductor section area must be suitable to limit operating temperatures and be fitted with a protective sheath of suitable color for safe identification of conducting function (neutral, ground, phase, etc.).

Protective measures against direct and indirect contact — Using 25 V AC and 60 V DC electricity is efficient protection against direct contact. At such low Voltages conducting elements can be exposed. At low voltages, active parts of circuits need neither be grounded nor linked to the protective conduction system of other circuits. In fact, this could cause short circuiting with consequent dangerous rise in voltage in circuits where users normally take no particular safety precautions.

1. Passive Protection System

Protection through insulation, sheathing and barriers — The live parts of high tension wires must be completely and adequately insulated and grounded. The insulator must be resistant to mechanical, thermal, and chemical (oxygen, nitrogen, ultraviolet rays) damages. Electrical equipment must be contained in elements which ensure protection against direct contact and all types of mechanical damage. These elements are "sheaths". "Barriers" are elements which ensure against direct contact by prohibiting access. To ensure absolute safety, sheaths and barriers must be fixed and removable only with the help of specific tools.

Reinforced insulation — This means using basic insulation plus reinforced insulation. If the basic insulation should give way, the second layer ensures protection. Equipment with double insulation must be marked with the symbol of Figure 20. Further protection is offered by using the equipment in insulated, grounded areas with no extraneous metal masses in the environment. Loss of basic insulation is not dangerous to individuals in this case. Machinery can also be run by independent dynamos or electrical sources isolated from other electrical circuits and from the ground (isolated transformer). In case of basic insulation breakdown, this protective measure works where current capacity is low, i.e., on limited secondary circuits.

2. Active Protection System

Aside from systems which impede contact, there are others which limit to safety levels the current that passes through the body in case of direct contact. A differential switch which

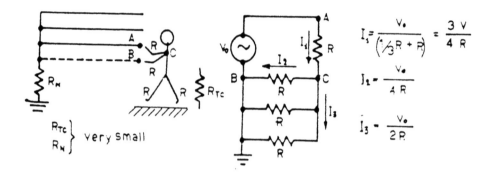

FIGURE 21.   Example of bipolar contact (neutral phase) and representation of the equivalent circuit.

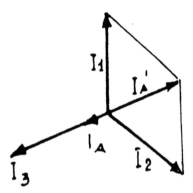

FIGURE 22.   Vectorial diagram of current in triphasic system where there is dispersion current in two phases.

has "physiological sensitivity" is such a measure. Its current when the switch is "on" is

$$I < 30 \text{ mA} \tag{22}$$

This value, if limited to brief exposure times, is a compromise between the need to protect people and that of keeping the machinery working (see Figure 15).

Under special conditions efficiency of differential circuits is damaged. When bipolar contact (neutral phase) takes place, the differential switch does not intervene if the person is insulated from the ground and may not intervene if the person is not insulated from the ground (see Figure 21).

In the first case all the current that comes out of "A" goes into "B" and since there is no difference between the current in the two branches of the circuit, the differential switch does not come into play.

In the second case, the differential comes into play only when the difference is between I_3 and I_1 and the protection is, therefore, partial.

In a triphasic system where there is dispersion current during two phases (e.g., 20 mA) the differential switch will be sensitive to their vectorial sum (which is 20 mA).

If a person touches the third phase sending a 30-mA current towards the ground, the differential switch will register only the vectorial sum of the currents which is 20 mA and will not come into play (Figure 22).

Shortcircuiting of one of the wires is another problem limiting the effectiveness of high sensitivity differential switches. In this case, the current flowing through the body is decreased by the flux generated by current induced in the shortcircuited wiring. This current, according to Lenz's law, is in opposition to the current that induced it. This is the situation when using a star connection with central ground where there are grounded neutral lines downstream from the differential switch.

A situation similar to dispersed current holds for continuous components (e.g., rectifiers) running on unbalanced triphasic current.

Though some of the situations described above are highly improbable, the differential switch must be considered only as an ulterior safety measure to be added to passive protective measures.

Protection against indirect contacts — Protection against indirect contacts is achieved by attaching machinery to a grounding system using special conductors called "protective conductors". If the machinery should be charged, protective systems click in, limiting the overload on the electrical system, calibrated in harmony with the safety curve. If ground resistance is Rt and the protective system click in within 5 sec, current is Ii, then:

$$Rt \leq 50/Ii \tag{23}$$

The protective measures mentioned can be composed of automatic switches, fuses or differential switches. In this case, I takes the place of Ii.

Degree of electrical installation protection — The distinctive features of the environment where electrical installations are located determine the characteristics of sheaths and barriers suitable to ward off all possible dangers. The features are identified by "degree of protection" and marked, by the IEC, with the letters IP followed by two numbers. The first number indicates protection against introduction of solids; the second indicates protection against liquids.

The first number can range from 1 (a 50-mm diameter sphere cannot pass through the sheath) to 6 (talcum powder sufficient to leave a visible deposit cannot filter through).

The second number can range from 0 (no protection) to 8 (protected even in immersion for an unlimited period of time).

Electrical installations in high risk areas — In some areas, electrical installations must be set up using special precautions because of the high risks run by workers. This holds in the following cases:

- Areas where there are limited size conducting surfaces (metal containers etc.)
- Medical rooms
- Bathroom and showers
- Swimming pools
- Areas where there is fire or explosion risk
- Construction sites

Given the complexity of the discussion for each situation please consult specific literature for safety measures. Here we will give only essential information on electrical installations for constructions sites and arc welding.

Construction sites — Construction sites are high electrical risk areas because installations are temporary and are continuously modified as building progresses. They are also called on to do heavy work and are used by many different more or less competent people making systematic control difficult. Therefore, precautionary measures must be taken to protect wiring both from solids and from liquids. Protective measures keeping solids over 2.5 mm out and rendering installations waterproof (e.g., against pouring rain) are recommended. Switchboxes must have interblocking controls impeding opening, and live parts must be accessible only when the boxes are open. Emergency shut-off controls and key switches must be found on all equipment which could be dangerous if turned on at the wrong moment. The relationship between protection and ground resistance should be 25V instead of 50V, so Formula 23 becomes:

$$Rt \leq 25/Ii \qquad (24)$$

There must be overall use of high-sensitivity differential switches. A severe risk is contact between machinery or construction equipment and aerial or underground wires. If the wires are less than 5 ft from construction equipment, they must be moved or turned off. If the wires carry low voltage, the sector involved may be isolated or the area for machinery can be marked off, or fixed barriers or portal signs can be set up.

D. Arc Welding

In arc welders step-up tension may reach 80 V AC, because if it drops below safety levels the arc voltage would not be produced. Thus, the welder touching the electrode and the return cable is subject to dangerous voltages. In these cases, given the decreased current values, personal protection measures (leather gloves) are sufficient for safety. The situation is much more dangerous when (due to loss of insulation) the primary circuit short circuits with the secondary circuit. The welder could be subjected to all of the primary circuit current without the breakdown being noticed. This is why the poles of the secondary circuit should be isolated from the ground and the two windings separated by double insulation and a screen linked to the ground.

When welding is carried out with the welder in close contact with the piece to be welded (e.g., in a metal container) the equipment must not touch the piece. The electrode pliers must be completely covered with insulation and the welder must use precautions such as insulated foot boards, etc.

If a series of welders is fed from a single triphasic source, the electrodes of the different machines should not be accessed at the same time because the current is linked among them.

If a series of monophasic machines is used on the same piece, the current between two electrodes may be null or double according to the way they are set up.

Given its importance, wandering welding current needs a section to itself; it is present when the welding current, which may be on the order of hundreds of ampheres, does not go through the return cable but goes elsewhere. This current may involve structures far from the welding station and cause explosions, fires, or breakdown of mechanical parts or isolation of electrical conductors.

E. Socket Safety Requisites for Plugs

Plugs must in no way leave the possibility of direct contact open when the plug is completely or even partially inserted in the socket. The grounding prong must be the first to go into the socket and the last to leave it. Only if the plug is taken apart can the prongs be removed and the openings must be elastic enough to ensure good contact. Clamps must clasp the conductors between metal surfaces and must be placed so as to impede contact with the protective conductor. The flexible cable must be secured to the plug using a plug-cable system that ensures low mechanical wear on the conductor endings. If the installation is made up of systems using different voltages, the plugs must not be interchangeable.

F. Rooms for Storage Batteries

Dangers linked to storage batteries, aside from electricity itself, are contact with solid reagents and/or liquid electrolytes and gases produced during charging. Especially large fixed batteries (150 V and 150 KAh for a total of 8 h) should be placed in specially designed rooms and protected from humidity, infiltration of gases and vapors from chemical industries, stables, cesspools, and distilleries not to mention dust and vibrations.

The precautions are especially important for lead batteries in open containers. These rooms must be well ventilated and large enough to ward off formation of explosive gases when hydrogen is produced during charging. Remember that the explosion point for hydrogen is between 4.1 and 72.4% in the air and that almost all the difference between the charged in (Ah) and the discharged state is due to the electrolysis of water. In quantitative terms, that means that for every Ah lost through electrolysis, 0.6212 of explosive gas are produced. Therefore, rooms where storage batteries are kept must ensure enough ventilation to keep the concentation of explosive gases within 1.1%.

Lead-acid storage batteries in open containers require ventilation equal to P (cu.m/min)

$$P \geq \frac{Q \cdot n}{1000 \cdot t} \tag{25}$$

where Q = battery capacity in Ah, n = number of cells, and t = total duration of charged state in hours.

Alkaline or closed-container lead-acid batteries with filtering lids need minimum cubature C (m^3)

$$C = \frac{Q \cdot V}{100} \tag{26}$$

where V = nominal battery voltage.

It is also important to remember when planning ventilation that battery room temperature should never be more than 40°C nor less than 10°C.

When forced ventilation is necessary, the electrical wiring of the motor must be of the antideflagration type and placed so as not to be touched by the explosive gases. Air from battery storage rooms must absolutely not flow into other closed areas. The rooms exclusively for batteries should measure 2 m from floor to ceiling and both floor and walls should be covered with dustproof-acid-resistant materials. The floor should be sloped and have drainage outlets. Windows should be protected with anticorrosive screens to keep out any foreign objects.

For stability reasons the battery room should be on the ground floor. No water, gas, or heating pipes should cross the room and nothing but batteries should be stored there.

First aid and emergency stations for personnel should be located near the clearly marked entrance. These stations should have eyedrops and neutralizing solutions such as bicarbonate of soda, boric acid, etc.

The following signs should be on the door: "danger of explosion and corrosion", "entrance forbidden to unauthorized personnel", and "individual protection measures obligatory". The meaning of conductor colors with respect to polarity and current, caustic ation of electrolytes (sulfuric acid and potassium hydrate), and of lead dust or fumes should also be clearly marked.

Batteries must be installed so that the number of cells connected gives a final charge of no more than 220 V with a special mechanism to ward off any possible contact between parts with a difference in potential of over 220 V and personnel.

Corridors 70 cm wide must be left open for easy upkeep (Figure 23).

Extra care must be taken to avoid personnel accidently grounding batteries. The batteries themselves must be surrounded by insulation footboards whenever there is tension greater than 220 V. It is best to insulate cells from each other and from the ground whenever their value in ohms is no less than 1000 times their nominal voltage. Joints must ward off any possibility of sparks and be protected from corrosion. All conductors must be painted in colors that identify them precisely to protect them from corrosion.

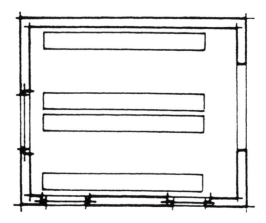

FIGURE 23.   Typical room for battery storage.

VII. ELECTROMAGNETIC FIELDS AND MICROWAVES

A. Effects on Living Organisms

At present not much is known about the effects of electromagnetic fields on cells of living organisms. At the macroscopic level, exposure to sufficiently strong electromagnetic fields or microwaves induces a rise in tissue temperature proportional to intensity and duration of exposure. For this reason some Western nations have proposed standards fixing exposure limits for these physical agents.

Easterm European reseachers have found that even electromagnetic fields too weak to induce temperature changes do have other effects on living tissues and have therefore proposed notably lower limits.

B. Measurement of Electromagnetic Fields

Only "distant" or Fraunhofer fields are relatively easy to measure since they are at least a few wavelengths from the source.

In these zones application of the classical theory of electromagnetism yields precise evaluation of the power emitted moving orthogonally with respect to the electrical field vector $\vec{E}$ and magnetic field vector $\vec{H}$, over a continuous area perpendicular to the direction of propagation. This value, called power density or intensity, is given by the formula:

$$Ps = 1/2 \ E^2/Zo = 1/2 \ Zo|H|^2 \tag{27}$$

This formula does not hold for "near fields".

The induction zone is of fundamental importance in accident prevention since, in almost all cases, workers exposed to this risk work near the source. In fact, if short distances are sufficient for "distant field" conditions when dealing with microwaves, distances range from a few meters to a few hundred meters when dealing with radiofrequency waves.

In general, distance from the source at which the field can begin to be considered puntiform is given by:

$$2 \ D^2/\lambda \tag{28}$$

where D = maximum size of the source and λ = wave length.

Instruments that measure only the electrical component in V/m (volt/meter) or the magnetic component in A/m (ampere/meter) in a "near field" yield data only on apparent power density or Ps (formula 26) and not on wave intensity. This is obtained from Poynting's vector:

$$S = (\vec{E}\wedge\vec{H}) \tag{29}$$

From the above it is evident that electromagnetic fields are characterized by intensity of electrical and magnetic fields in the induction zone (near field) and by the intensity of the electromagnetic field in the radiation zone (distand field).

In the induction zone, the field is mostly electrical so its measurement can be useful (given that the sounding device used does not pick up interference from the magnetic field rendering results meaningless). Remember that, at the start, there is no way to know which component is more intense nor how much more. The measuring device is made up of an antenna sensitive to the magnitudes to be measured, a sounding device that converts AC induced in the antenna into DC, and an instrument that measures the magnitude of direct current and expresses this in terms of magnitude of the field being measured. This type of apparatus called a "direct detector" gives the RMS value of the field and the mean density of power. There are more sophisticated apparatus but they are not suitable for "on the job" measurements.

Antennas must be small referring to wavelength if they are to measure only one of the components of the field. In fact, they must be comparable to lumped constant circuits. When antennas are comparable in size to wave lengths they are like distributed constant circuits and therefore a source of electromagnetic fields. If there is an open circuit structure the tension will be significant but the current will be very low and the electrical field will prevail. If the circuit is a closed type, a magnetic field will prevail.

C. System to Decrease Risk of Exposure to Electromagnetic Field

Exposure to electromagnetic fields can be decreased in various ways by acting on the source, the person exposed, or by decreasing work time in the presence of electromagnetic fields.

The most incisive action is that taken on the dynamos themselves especially if measures are built in during the planning phase. Usually precautions are taken in all three areas.

Precautions taken are grounded shields connected to the body of the machinery and barring off of zones where there are electromagnetic fields.

Insulation of machinery from the electrical feeding network by means of network filters is another common precaution.

1. Metal Shields

It is well known that metal shields impede transmission of electromagnetic fields both by reflecting them off both sides as they pass and by the dispersion of energy inside the shield. The screening action is based on the characteristics of electromagnetic fields (and especially on frequency), the metal used to make the shield, and its thickness.

For distant fields (E/H = 377 Ω) the attenuation found using aluminum or iron shields is shown in Figure 24.

Problems exist, as you can see, only for relatively low frequencies (100 KHz) though attenuation is greater than 10. In the induction zone, attenuation depends on how far the source is from the shields. This varies for electrical fields and magnetic fields.

Whereas the electrical field poses no problems because it can always be heavily attenuated, the magnetic field becomes more difficult to attenuate as frequency and distance between the source and the shield decrease.

In static fields, and whenever shields are made of nonmagnetic materials, attenuation goes to 0.

Significantly attenuating shields at values lower than a few KHz is extremely difficult.

Figures 25 and 26 show the trend in attenuation as a function of frequency for screens 10 cm from the source and for steel under the same conditions, as a function of thickness.

It is clear that under near field conditions at low frequencies, screens make sense only if constructed in very thick magnetic materials.

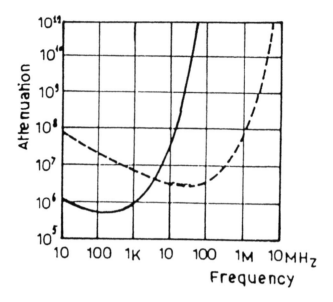

FIGURE 24. Attenuation of distant fields obtained using aluminum or iron shields; ——— steel 0.6 mm; ---- aluminum 0.5 mm.

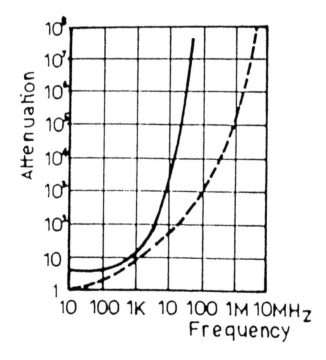

FIGURE 25. Attenuation as a function of frequency obtained with steel or aluminum screen placed at a distance of 10 cm from the source; ——— steel 0.5 mm; ---- aluminum 0.5 mm.

As screen distance from the source increases the magnitude of the magnetic field, with respect to that of the electrical field, decreases and attenuation increases.

In general, total screening attenuation is (frequency of a few megahertz) so great that it brings the intensity of all dispersed fields well within the lowest admissible values.

Total screening involves such notable technical difficulties, that it, is not possible under

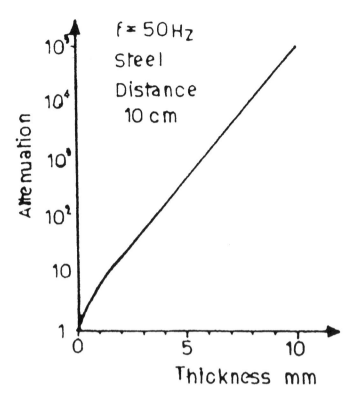

FIGURE 26. Attenuation as a function of thickness obtained with a steel screen placed at a distance of 10 cm from the source.

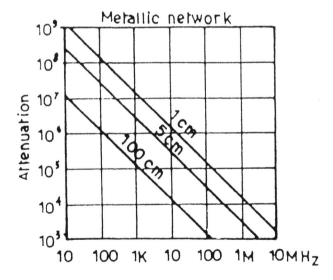

FIGURE 27. Attenuation obtained using a discontinuous screen realized with a metallic network.

normal conditions, because it is both impractical and very costly. The tendency is to extend screening as the problem increases with discontinuous screens instead of continuous ones. This solution is fine for electrical fields but inadequate for magnetic fields especially at low frequencies and at a short distance from the source. Figure 27 shows attenuation using a metal net discontinuous screen.

It is important to remember that the openings in the screen can give rise to restricted fields more intense than those found where no screen is present. The openings must therefore be constructed so as to ensure good electrical contact at least every 1/20, better if every 1/100.

The feeding or measurement cables (shielded) must have their shielding connected to the main screen, where they cross it.

It is also important to install filter or separator transformers along the cables.

The concept of the matallic screen can also be applied to workers using Faraday boxes or having them wear lattice work fabric suits with magnetic wiring connected to the ground and to the mass by an "umbilical cord". If there is very high frequency partial shields can be used because where there are very small wavelengths border effects become negligible.

By connecting to the ground or the mass we mean making a sound connection with the field frequency in question. In fact radiofrequencies differ from network frequencies, where a small copper section is sufficient, in that, given the "skin effect", good connections can only be had using short conductors of ample perimeter placed every 1/20.

It is better to place the apparatus on a sheet of metal.

2. Filters on Adduction Cables

Since large RF quantity leaks are had on passage through metallic structures, it is advisable to install conductor cut-off controls for RFs characterized by low-pass filters. To this end the following are used:

- An isolation transformer with an overall shield and a shield between the primary and secondary circuits
- A neutralizing transformer with functions like those of the isolation transformer; it is made up of two like windings in series with a connected conductor
- Network filters on each conductor; these are passive reactive components (inductance and capacity) which allow low frequency current flow but high impedance to radio frequencies.

The set ups generally used are shown in Figure 28.

3. Segregation of Dangerous Zones

In relatively short wave fields it may be convenient to isolate only those zones where the field is especially intense so that personnel can enter them only in cases of absolute necessity.

The risk of electrostatic charges on nonconducting materials caused by the electromagnetic fields still remains.

Barriers must be in conducting materials with high-resistance grounding to allow gradual discharge towards the ground even when the workers themselves are electrostatically charged.

If contact were direct, the person discharging towards the ground would experience the same unpleasant feeling as he would if he touched a surface charged to the same degree. It is therefore advisable to lay tiles having 4000 to 5000 Ω/m^2 conductivity (like those used in operating rooms) on the floor of rooms where electromagnetic field-generating machinery is located.

The tiles must be laid using electroconductor glues.

A conducting network in copper strips, interwoven closely enough to take in every tile, must be placed between the groundwork and the tiles themselves.

Workers with pacemakers must absolutely avoid entering zones where electromagnetic fields are active. This must be clearly signaled through safety signs.

VIII. LASER

Except for low power laser beams used for aligning larger machinery, laser generators

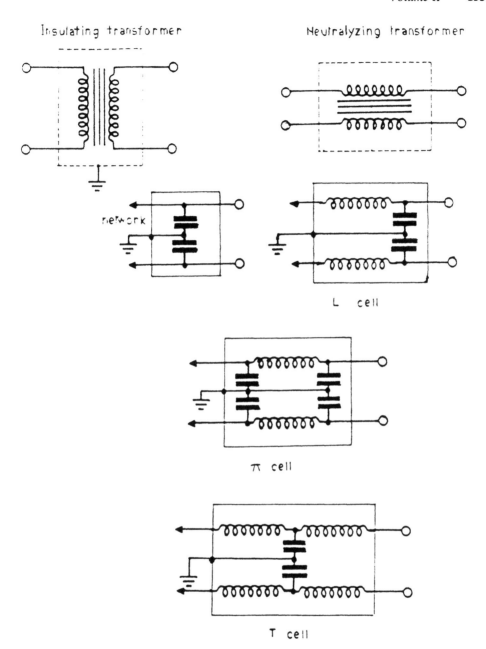

FIGURE 28. Filters on adduction cables.

should be run so as to avoid all contact between beams and parts of the body, especially the eyes.

The entire course of the beam should be isolated using continuous or discontinuous slot shields when machinery is placed along it (see Figure 29).

Shields must be made of materials resistant to the laser activity.

Discontinuous shields should have flat black surfaces. Shields must be connected to each other and to the laser generator by microswitches so that removal of any part triggers a diaphragm blocking off the beam.

This precautionary measure must not be automatically reversible. The worker must per-

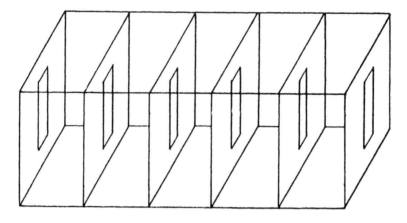

FIGURE 29.   Discontinuous slot shields for laser.

sonally switch off the diaphragm and prepare it for future action thus taking note of the breakdown in the shielding.

The optical plane must not have reflecting surfaces (best if flat black) and the walls should be covered with strips of nonreflecting material where beam impact is most probable.

All other surfaces must diffuse light.

Frosted glass windows must be impervious to laser light.

The power-on switch of the generator must also turn on a light on the laser source and another flashing light with "laser in action" written on it over the entrance to the room.

When powerful beams are generated, opening the door to the laser room should set off an acoustical alarm and trigger a diaphragm over the source through a series of micro switches. These protective devices and reset must be manually shut off and reset.

The target zone must be totally barred off. All materials and furniture which could possibly be touched by beams must be nonflammable. When the target zone is in a different room from the laser source, this room must be protected as above and, if personnel is present, opening of the diaphragm or turning on of machinery must be possible from this target room. Adequate personal protective devices must be distributed as well.

IX. NOISE

Working area noise levels must ensure "reasonable" safety from the risk of hearing loss due to environmental noise. Reasonable safety means that only a nonsignificant percentage of workers exposed will suffer hearing impairment after having worked 8 h/d, 5 d/week for 20 years.

As of 1987 European Economic Community authorities have established 85 dBA with a $q = 3$ exchange factor as the noise limit in work areas. These values, aside from local variations, are held to be the safest by authorities all over the world.

A. Noise Evaluation

Noise is measured as the level of acoustic pressure P as compared to a 20 μ Pa sound pressure P_o:

$$L = 20 \lg P/P_o (dB) \qquad (30)$$

Since ear sensitivity varies for different frequencies (i.e., at equal pressure sounds are perceived as more or less intense according to their frequency), the sounding device should

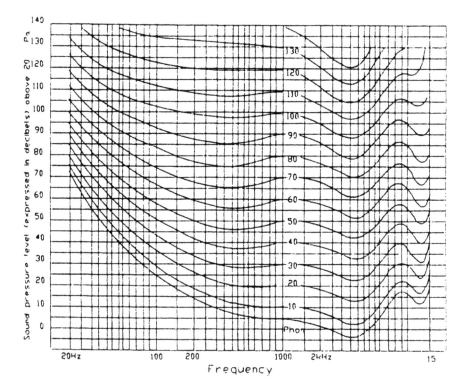

FIGURE 30. Free-field equal-landness countours for pure tones (ISO/R226).

be equipped with an electrical ponderation network that feeds the measuring device data analogous to the isophonic curve of the audiogram shown in Figure 30.

Given the complexity of the problem three ponderation curves, A, B, and C, were chosen corresponding the isophonic curve at 40, 70, and 100 phon. Experience has shown that using three scales creates many problems expecially for border values of each curve. Curve "A" was chosen for this reason even for very high values.

As for variations in level, the most recent laws have defined an equivalent noise level which is that continuous noise level which yields the same energy effects as the sum of variable sound levels. Equivalent levels are defined by the formula:

$$\text{Leq} = \frac{10Q}{3} \log\left[\frac{1}{T} \int_0^T 10^{3L/10Q}\right] dt \tag{31}$$

where T = total exposure time, L = acoustic pressure level, and q = exchange factor.

According to the law, either 3 or 5 is assumed as change factor value. The factor value 3 is chosen when the hypothesis is that a noise of given sound level for single intervals of time T yields the same Leq values as halving the single times T and adding 3 dBA to each level. In other words, the weighted energy constant principle is held as valid.

When Q = 5 dBA then values for increase in level must be such that the product of S^2T, where S is sound perception and T is time, remains constant.

B. Ways of Decreasing Environmental Noise

First of all noise sources must be decreased. Compressed air outlets should be silenced and gear transmissions should be cushioned (a straight-toothed gear is noisier than one with helical teeth).

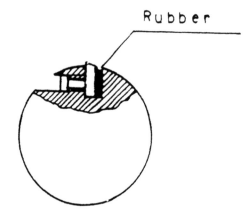

FIGURE 31.   Example of current tool of combination planer.

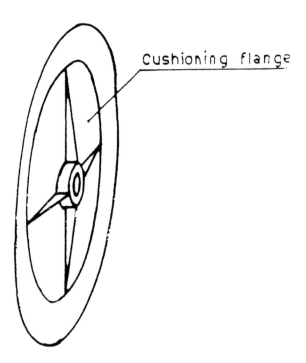

FIGURE 32.   Disk of rubber material for damping noise in bench circular saw.

Woodworking machinery tools must not be hollow (combination planer see Figure 31) or fixed damping flanges (bench circular saw, see Figure 32).

Air outlets must be bidirectional and have connecting conduits so as to avoid areas of turbulence.

Machines must be installed with antivibration junctions so there is high mechanical impedance to transmission of vibrations to building structure.

Noisy machinery and production phases should be acoustically isolated from other production lines.

Where there is high circulating energy baffles should be installed to decrease it.

Given an environment of specific architectural and interior design, it is possible to predict acoustic pressure after installation of a certain quantity of machines of known sound power production (it is measured using dual channel FFT narrow band analyzers and then evaluated electronically).

Environmental absorbtion is found measuring mean echo time.

Acoustic pressure levels are found using the formula:

$$10 \lg W/Wo = 20 \lg P/Po - 10 \lg T/To + 10 \lg V/Vo - 14$$

The result is in dB.

Therefore,

$$Lw = Lp - 10 \lg T + 10 \lg V - 14(dB) \tag{32}$$

where Lw = sound power level, L = acoustic pressure level, T = echo time, V = environmental volume, and 14dE = logarhithmic expression of environmental constants.

If equivalent acoustic pressure is still over 85 dBA workers must be issued special protective devices and work in shifts.

Individual protective devices limit aural communication with the surroundings and this can be very risky in case of impending danger.

X. VIBRATIONS

When the body undergoes acceleration, dynamic forces proportional to m are brought to bear on ligaments inducing relevant movements of organs and skeleton with consequent friction and fatigue. This can lead to functional disease or, in extreme cases, lesions. The effects described depend on the frequency and the direction of vibrations.

In an erect or seated position at a vertical 2-Hz force, all parts of the body are subject to the same acceleration as the source transmitting motion. Over 2 Hz, acceleration on the heart and liver gradually increases, reaching a maximum between 4 and 8 Hz. This is, in fact, the range of maximum energy transfer deriving from better synchrony between the dynamic stimulus on organ mass and ligament elasticity. Resonance frequency is from 4 to 5 Hz for the heart and 6 to 8 for the liver.

Maximum oscillation of these organs at resonance frequency is limited by friction and by the presence of other organs (cushioning).

Resonance frequency for the head-neck-shoulders system is between 20 to 30 Hz. Vision is disturbed between 60 to 90 Hz (pupil resonance).

If stimulation occurs with the body in horizontal postion, the critical zone is between 1 to 3 Hz. This is due to transversal resonance of the head-neck-shoulders system (1.5 to 2 Hz) caused by a decrease in horizontal rigidity.

The situation is different for the limbs.

Hands are definitely subject to fatigue since the various parts of the hand holding vibrating tools undergo alternating compression cycles which, in vast movements, stimulate articulations to points between 2 to 3 Hz and 40 HZ. Smaller movements cause fatigue in nerve endings and blood vessels between 40 and 1000 Hz.

A. Evaluation of Vibrations

Research has shown that vibration effects depend not only on frequency but on intensity, direction, and exposure time as well. In other words, the human body is affected by the overall "dose" of vibrations. On a more or less linear scale this is a product of intensity and exposure time (see Equation 33).

$$dose = 100 \; ti/Ti$$

$$esp. \; equivalent = \sum_i (ti/Ti) * 100 \tag{33}$$

where ti = exposure time to Li (vibration intensity) and Ti = maximum allowed time for Li.

There are three main fields of research of evaluation of vibrations:

1. Transportation sickness up to 1 Hz (0.1 to 0.6 Hz)
2. Vibrations involving the whole body (1 to 80 Hz)
3. Vibrations involving the hand-arm system (8 to 1000 Hz)

(1) If we see the human body as a linear system controlled by weight alone, then we can assume that reaction is proportional to the product of (a·t). Doubling intensity produces the same effects in half the time (linear relationship).

(2) The mechanical response of the body is not linear when the range is from 1 to 80 Hz (resonance effect). Fatigue brings "exposure level" curves into play which show greater effects for longer priods of time.

(3) This area is even more complex. However, the above considerations still hold true. The exposure level curve is divided into time intervals in order to utilize variations due to regular breaks from exposure.

Figure 33 illustrates the above points.

It is clear that standards set to limit bodily exposure do not allow for any values over 1 to 9 peak since the body is automatically distanced from any surface producing more than that stimulus value.

To avoid damage to structures standards have been proposed (Figure 34) defining the upper and lower limits for acceleration of vibratory movement.

International standards are still in the proposal phase.

B. Technical Measures to Limit Exposure to Vibrations

Measures to decrease entity of vibrations can be divided into action taken on the sources or action taken on installation of machinery causing vibrations.

In the first instance the masses in rotation are better balanced and alternating movement masses are decreased.

As far as installations are concerned, mechanical impedence of the system must be increased to decrease energy transfer through solids. This means that vibrating movement be mechanically isolated from the rest of the structure.

XI. ULTRA-HIGH-FREQUENCY SOUNDS

These are mechanical vibrations having frequencies over 20 KHz; therefore, beyond the hearing threshold they are produced by piezoelectricity.

It is possible to produce ultrasounds with frequencies up to 15 MHz and, when working on the upper harmonics, up to 200 MHz. Specially shaped crystals permit focalizing effects.

A. Safety Limits

At the moment it seems that for exposures under 100 W/cm^2 no adverse biological effects are felt.

For short periods of time the following values are considered safe (given they do not go beyond 50 J/cm^2):

50W/cm^2 for 1 s
5w/cm^2 for 10 s
1W/cm^2 for 50 s
100 mW/cm^2 for 500 s

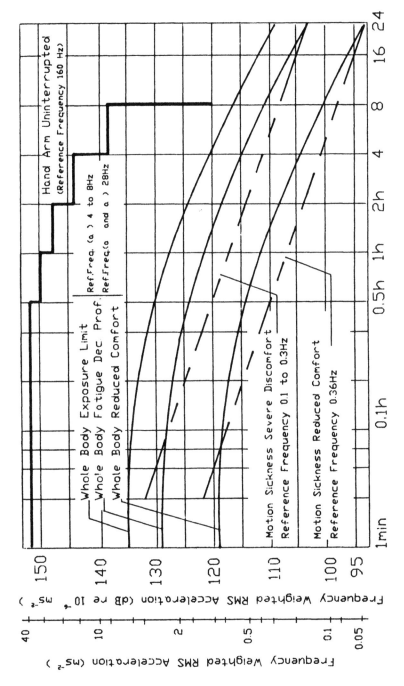

FIGURE 33. Effect of vibration on human body: acceleration upper limit in vibrational motion.

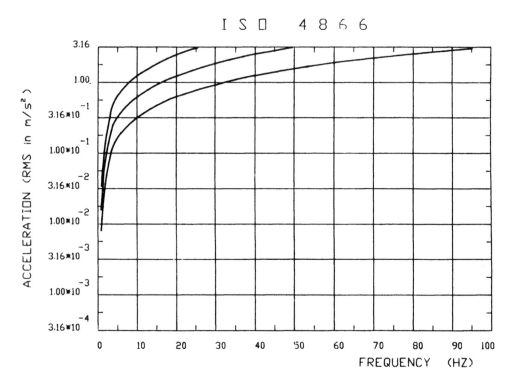

FIGURE 34.   Effects of vibrations on structures: acceleration upper and lower limit in vibrational motion.

Table 7

BIOLOGICAL EFFECTS FROM U.S. DATA AS A FUNCTION OF TOTAL EXPOSURE (VALUES UNDER 50 J/cm²)

Bibliography	Biological system	Exposure (J/cm²)	Effects
Prasad N. A.(1976)	HeLa cells	2.4	Decreased uptake of nucleic acid precursor
Sikov M. et al.(1976)	Rat	3.0	Retarded neuromuscolar development
Dyson M. et al.(1973)	Chicken embryo	3.0	Blood stasis
Sikov M. et al.(1976)	Rat	3.0	Retarded neuromuscolar reflex development
Anderson and Barrett(1977)	Mouse	3.0	Immune depression
Sankowiak S. et al.(1965)	Rabbit eye	6.0	Vascular mutation
Slovoto S. et al.(1976)	Root vicia faba	12.0	Chromosomic abberrations
Anderson and Barrett(1977)	Mouse	15.0	Immune depression
Williams A. R. et al.(1976)	Human blood	19.5	Decrease in ''in vitro'' coagulation time
Curto K. et al.(1976)	Mouse	22.5	Post partum mortality
Liebeskind et al.(1979)	HeLa	30.0	No DNA duplication
Tsutsumi Y. et al.(1964)	Dogs SNC	32.0	Rise in GOT in liquor

Note: Bibliographical data referred by Ciatti et al.

It is interesting to note the biological effects found experimentally using total exposure over or under 50 J/cm² (Tables 7 and 8).

Table 8
BIOLOGICAL EFFECTS FROM U.S. DATA AS A FUNCTION OF TOTAL EXPOSURE (VALUES UNDER 50 J/cm²)

Bibliography	Bilological system	Exposure J/cm²	Effects
Bender L. F. et al.(1954)	Dog bone	60	Hemorrhage
Galperin-Lemaitre et al.(1975)	Veal thymus and DNA salmon sperm	72	Degradation of DNA helix
O'Brien W. D.(1976)	Mouse	90	Decreased fetal weight
Galperin-Lemaitre et al.(1975)	Veal thymus	180	DNA degradation
Longo F. W. et al.(1976)	Rat	315	Reversible splenomegaly
Murai N. et al.(1975)	Rat	360	Altered emotional behavior, retarded neuromuscular reflex development
Shosi R. E. et al.(1976)	Mouse	720	Fetal anomolies
O'Brien W. D.(1976)	Mouse	900	Decreased fetal weight

Bibliographical data referred by Ciatti et al.

REFERENCES

1. **Paciucci, R., Petrella, L., and Carrara, D.,** Criteri construttivi dei laboratori di ricerca dal punto di vista della prevenzione infortuni e dell'igiene ambientale, Atti del Convegno Protezione e Sicurezza del Lavoro nei Laboratori Scientifici, Rome, February 8 to 10, 1984.
2. Standard per impianti ICI, England.
3. **Paciucci, R. and Petrella, L.,** Parametri che influenzano l'affaticamento visivo sul posto di lavoro con particolare riguardo ai videoterminali, Atti II Convegno Nazionale di Oftalmologia Sociale, Sorrento, May 2 to 5, 1984.
4. **Paciucci, R.,** Evaluation of illumination in the working environment in order to individualize factor which cause visual strain, 4th World Congress of Ergophthalmology, Eye Toxicology, Laser Safety, Lighting and Visual Well-Being in Living and Working environments, Sorrento, May 26 to 30, 1985.
5. **Clerici, C.,** Quaderno di elettrificazione, *Illuminotecnica,* Vol. 13, Delfino, Milano.
6. **Vallat, A.,** L'illuminazione fattore di produttività nell'impresa, CIS 105 b, 1960.
7. **Parvopassu, P.,** *Lezioni di Impianti Meccanici,* Sistema, Ed., 1963.
8. **Carrescia, V.,** Impianti di messa a terra, ENPI, 1974.
9. From Scientific American
10. **Grandolfo, Checcucci,** Millanta ed altri Protezione dai campi elettromagnetici non ionizzanti IROE, Florence, 1982.
11. **Campos Venuti, G., Grandolgo, M., and Mariutti, G.,** Ipotesi di normativa nel campo radiazioni a radiofrequenza e a microonde, January 31, 1979.
12. Manuale Misuratore Bruel & Kjaer Mod. 2512

Chapter 8

CHARACTERISTIC AND SAFE MANAGEMENT OF HAZARDOUS CHEMICAL SUBSTANCES (AS IN THE LIST OF CEE DIRECTIVE NO. 501)*

Claudia Bartolomei and Sergio Paribelli

TABLE OF CONTENTS

---

* The references in this chapter are the same as those that appear in Chapter 2, Volume I.

I. ACETYLENE

CAS RN 74862, CEE q: 50 t.

Formula — C_2H_2.

Flammability — Highly ignitable, particularly when exposed to heat on flame or by spontaneous chemical reaction flash point: $-17.8°C$ (closed cup). When ignited it burns with an intensely hot flame.

Explosion — Moderate explosion risk when exposed to heat or flame or by spontaneous chemical reaction. Explosion limits: 2.5 to 81%. Acetylene is highly instable at high pressures and even moderate temperature and may decompose into hydrogen and carbon with explosive violence if subjected to sparks, heat, or frictions.

Incompatibilities — It is incompatible with the copper, brass, copper salts, copper carbide, Hg, Hg salts, Ag, Ag salts, and halogens; heating or impact, in presence of these substances, may cause fires and explosions. Molten K ignites in C_2H_2 and then explodes. C_2H_2 reacts vigorously with trifluoromethyl hypofluorite. With O_2, C_2H_2 can detonate very powerfully. It also can react vigorously with oxidizing materials.

Toxicity — Acetylene acts as a narcotic (it has been used in anesthesia with 60% of oxygen) by diluting the O_2 in the air to a level which will not support life. In general industrial practice acetylene does not constitute a serious hazard. OSHA standard air: CL 2500 ppm. Occupational exposure to Acetylene recommended standard: CL: 2500 ppm.

Special precautions and preventive measures — Keep away from any possible source of ignition and combustible materials. Isolate from explosives, toxicants, radioactives materials, organic peroxides, chlorine, bromine, fluorine, copper, silver, and mercury. Outdoors or detached storage is preferred. Store in cool, well-ventilated places. Protect against lightning and static electricity. Piping used should be electrically bonded and grounded.

Disposal and waste treatment — "Fit a pipe line into a furnace or into a pit and burn with care."[2]

II. ACROLEIN

CAS RN 107028, CEE q: 200 t.

Formula — C_3H_4O.

Flammability — Dangerous when exposed to heat, flame, or oxidizers. Flash point: $-26°C$ closed cup and $-17.8°C$ open cup. Autoignition temperature: $235°C$ ($455°F$).

Explosion — Explosive limits 2.8 to 31 % by volume in air.

Incompatibilities — Contact with oxidizing agents, acids, alkalis and ammonia, amines, and metal salts, may cause fires and explosion; when heated to decomposition it emits highly toxic fumes (CO, peroxides).

Toxicity — Acrolein is highly toxic via oral, inhalation routes. Its main toxic effect is severe irritation of the eyes, skin, upper respiratory tract, and lungs, leading to pulmonary edema. It is a weak sensitizer. TCL_0: 300 ppb/2 h (administered to child via inhalation route;

pulmonary effects). LDL_0: 250 mg/kg (administered to man via intradermal route). OSHA standard: air: TWA = 0.1 ppm; LCL_0: 153 ppm/10 min (administered to man via inhalation route); TCL_0: 1 ppm (administered to man via inhalation route, irritant effects); TLV-TWA: 0.1 ppm (0.25 mg/m^3); and TVL-STEL: 0.3 ppm (0.8 mg/m^3). The time averaged TLV of 0.1 ppm is sufficiently low to minimize, but not entirely prevent, irritation to all exposed individuals.

Special precautions and preventive measures — Acrolein should be handled in a fume hood or in a closed system with exhaust ventilation of adequate scrubbing facilities. It should be stored in air-tight containers filled with nitrogen gas and located in a cool, well-ventilated place, away from any possible source of ignition or fire. Alkalis, acids, or oxidants should not be stored nearby. Outdoor isolated storage is preferred. Provide an adequate ventilation. Vapor-tight chemical goggles, chemical cartridge, or air line respirators, rubber gloves, apron, and boots, should be employed.

Disposal and waste treatment — Dissolve in a combustible solvent, then spray the solution into the furnace with after burner. "Acrolein may also be disposed of by absorbing in vermiculite, dry sand, earth, or a similar material and disposing in a secured sanitary landfill."[3]

III. ACRYLONITRILE

CAS RN 107131, CEE q: 200 t.

Formula — C_3H_3N.

Flammability — Acrylonitrile is highly ignitable and flammable, particularly when exposed to heat, flame, or oxidizers. Flash point: 0°C (open cup). Autoignition temperature: 481°C.

Explosion — Explosive range: 3.05 to 17% by volume in air at 25°C. Moderate explosion hazard when exposed to flame.

Hazardous potentials — Acrylonitrile violently polymerizes in the presence of concentrated alkalis; pure acrylonitrile tends to polymerize even at room temperature by light. It can react vigorously with oxidizing materials. On contact with acid, acid fumes, water, or steam, they will produce toxic and flammable vapors.

Toxicity — Acrylonitrile is an irritant to skin and eyes. It may affect the blood cells and the central nervous system. It is highly toxic via oral, inhalation, dermal routes; it is a suspected human brain carcinogen substance. By inhibiting the respiratory enzymes of tissue, it renders the tissue cells incapable of oxygen absorption. TLV-TWA: 2 ppm (skin). OSHA standard air TWA: 2 ppm. CL: 10 ppm/15 min. Occupational exposure to acrylonitrile recommended standard: CL: 4 ppm. The standard states that the employer shall assure that no employee is exposed to skin contact or eye contact with liquid acrylonitrile.

Special precautions and preventive measures — Separate from any sources of ignition and combustible materials. Outdoor or detached storage is preferred. For indoor storage, standard combustible materials storage room should be used. Protect from alkalis and oxidizing agents. Protect containers against physical damage. Provide an adequate ventilation, chemical cartridge respirator, or oxygen masks. Wear rubber gloves and aprons.

Proper decontamination support system, oxygen, antidotes, trained first aid personnel, and transportation methods should be immediately available.

Disposal and waste treatment — "Add by stirring excessive alcoholic sodium hydroxide. After 1 h evaporate alcohol and add sufficient calcium hypochlorite. After 24 h drain into the sewer with abundant water."[2]

IV. ALDICARBE (CARBANOLATE)

CAS RN 116063, CEE q: 100 kg.

Formula — $C_7H_{14}N_2O_2S$.

Hazardous potentials — When heated to decomposition it emits very toxic fumes of NO_x and SO_x.

Toxicity — Highly toxic via oral, subcutaneous, and dermal routes. It is a systemic poison, pesticide, reumatocide, acaricide.

LD_{50}: 900 mg/kg (administered to rat via oral route).

LD_{50}: 2500 μg/kg (administered by skin).

LD_{50}: 300 μg/kg (administered to mouse via oral route).

LD_{50}: 1400 mg/kg (administered to rabbit by skin).

LD_{50}: 2400 mg/kg (administered to guinea pig by skin).

LD_{50}: 3400 μg/kg (administered to duck via oral route).

V. ALLYL ALCOHOL

CAS RN 107186, CEE q: 200 t.

Formula — C_3H_6O.

Flammability — Ignitable and combustible. Flash point: 21.11 (open cup) and 23.89 (closed cup). Ignition temperature: 378°C. Flammable limits: 2.5 to 18% by volume in air.

Explosion — Vapor forms explosive mixtures with air.

Hazardous potentials — When heated it emits toxic gases and vapors (such as CO).

Incompatibilities — Contact with strong oxidizers may cause fires and explosions.

Toxicity — Allyl alcohol is highly toxic via oral, inhalation, intraperitoneal, and dermal routes. The vapor is highly irritating to the eyes and nose; skin absorption produces injury to the liver and kidneys in animals; skin contact causes burns and absorption through the skin; it may result in pain in the underlying tissue and gastrointestinal symptoms; contamination of the eye with the liquid produces corneal damage. TLV-TWA: 2 ppm (skin); TLV-STEL: 4 ppm; TCL_0: 25 ppm (irritant effects); and OSHA standard: air TWA 2 ppm (skin) averaged over an 8-h work shift. The time averaged TLV of 2ppm and the STEL of 4 ppm would appear to provide protection against systemic effects and injury to superficial areas of the body, and to provide a reasonable freedom for most individuals from irritation.

Special precautions and preventive measures — Allyl alcohol will attack some forms of plastics, rubber, and coatings. Separate from sources of ignition. Outdoors or detached storage is preferred. Indoor storage should be in a standard flammable liquid storage room. On standing for several years this chemical will polymerize into viscous syrup like polymer. Provide an adequate ventilation. Wear chemical goggles or face shield, chemical cartridge respirator, and other appropriate protective clothing necessary to prevent any possibility of skin contact with liquid alcohol.

Disposal and waste treatment — "Spray into a furnace. Incineration will become easier by mixing with a more flammable solvent."[2] "Absorb in vermiculite dry sand, earth, or a similar material and disposing in a secured sanitary landfill."[3]

VI. ALLYLAMINE

CAS RN 107119, CEE q: 200 t.

Formula — High fire hazard when allylamine is exposed to heat or flame; highly reactive also.

Flammability — Flash point: −28.89°C. Ignition temperature: 374°C.

Explosion — Explosive range: 2.2 to 22%.

Hazardous potentials — Dangerous, when heated to decomposition, it emits toxic fumes of NO_x; it can react with oxidizing materials.

Toxicity — Allylamine is a high human irritant substance via oral, inhalation, dermal routes. TCL_0: 5 ppm/5mn (via inhalation route; pulmonary effects).

Special precautions and preventive measures — Keep containers closed. Protect against physical damage. Outdoor or detached storage is preferred. For indoor storage use standard combustible liquid storeroom. Separate from oxidizing and combustible materials. Wear butylia rubber gloves, a face shield or a multipurpose gas mask, and coveralls. Advise extraordinary precautions against fumes.

Disposal and waste treatment — "Dissolve in a combustible solvent such as alcohol. Burn in an open furnace igniting from a safe distance with the utmost care or sprinkle into the fire chamber of the furnace with afterburner and scrubber, or pour into sodium bisulfate in a large evaporating dish. Sprinkle water and neutralize drain into a sewer with sufficient water."[2]

VII. 4-AMINOBIPHENYL (4-BIPHENYLAMINE)

CAS RN 92671, CEE q: 1 kg.
Formula — $C_{12}H_{11}N$.
Flammability — Moderate fire hazard when exposed to heat, flames, sparks, or powerful oxidizers. Autoignition temperature: 450°C (842°F).
Hazardous potentials — When strongly heated it emits toxic fumes.
Toxicity — Carcinogenic determination: positive (IARC). It is highly toxic via oral and inhalation routes. It is an irritant substance. It has caused bladder cancer in humans and experimentation animals. Effects resemble those of benzidine.
TDL_0: 4560 mg/kg (administered to rat via subcutaneous route, over a period of 44 weeks, intermittently. Equivocal tumorigenic agent.
TDL_0: 1520 mg/kg (administered to mouse via oral route, over a period of 39 weeks, intermittently; carcinogenic effects).
TDL_0: 660 mg/kg (administered to dog via oral route, over a period of 3 years, intermittently; equivocal tumorigenic agent).
TD: 5000 mg/kg (administered to rat via subcutaneous route, over a week intermittently; equivocal tumorigenic agent).
TD: 5580 mg/kg (administered to dog via oral route, over a period of 4 years, intermittently; equivocal tumorigenic agent).
LD_{50}: 500 mg/kg (administered to rat via oral route).
LDL_0: 250 mg/kg (administered to mouse via intraperitoneal route).
LDL_0: 25 mg/kg (administered to dog via oral route).
LD_{50}: 690 mg/kg (administered to rabbit via oral route).
Special precautions and preventive measures — Store away from any source of ignition (see Section XX).

VIII. AMITON

CAS RN 78535, CEE q: 1 kg.
Formula — $C_{10}H_{24}NO_3PS$.
Hazardous potentials — When heated via oral to decomposition it emits very toxic fumes of NO_x, PO_x, and SO_x.
Toxicity — Highly toxic via oral ipercutaneous routes. Ingestion may cause moderate erythema and moderate edema. It is a cholinesterase inhibitor.
LD_{50}: 5 mg/kg (administered to rat via oral route).
LD_{50}: 860 mg/kg (administered to mouse via oral route).
LD_{50}: 50 μg/kg (administered to mouse via intraperitoneal route).

IX. AMMONIA ANHIDROUS

CAS RN 7664417, CEE q: 500 t.

Formula — NH_3.

Flammability — Moderate fire hazard ignition temperature 651°C; flammable limits: 16 to 25%; the presence of oil or other combustible materials will increase fire hazard.

Explosion — Moderate hazard when exposed to flame or in fire. NH_3 + air in a fire can detonate.

Hazardous potentials — When exposed to heat, it emits toxic fumes of NH_3 and NO.

Incompatibilities — Contact with strong oxidizers may cause fires and explosions. Contact with calcium, hypochlorite bleaches, gold, mercury, and silver may form a highly explosive product. Contact with halogens may cause violent spattering.

Toxicity — Ammmonia affects the body if it is inhaled or if it comes in contact with the eyes or skin. It is also highly toxic via oral route; it is a powerful irritant of the eyes and respiratory tract. Contact with liquid ammonia may produce severe and skin burns. Preclude from exposure those individuals affected with eye and pulmonary diseases. Periodic surveillance is indicated. TLV-STEL: 35 ppm = 27 mg/m^3; TLV-TWA: 25 ppm = 18 mg/m^3; OSHA standard air: TWA 50 ppm; and occupational exposure to ammonia recommended standard air: CL 50 ppm/5 min.

LCL_0: 30000 ppm/5 min (via inhalation route).

TCL_0: 20 ppm (via inhalation route; irritant effects).

TCL_0: 1000 μg/kg (administered by skin; carcinogenic effects).

LCL_0: 10000 ppm/3 h (via inhalation route).

Special precautions and preventive measures — Most common metals are not affected by anhydrous ammonia; however when mixed with very little water vapor, both gaseous and liquid ammonia will attack vigorously copper, silver, zinc, other alloys, and some forms of plastics, rubber and coatings. Outdoors or detached storage is preferred. Indoor storage should be in a cool, well-ventilated, noncombustible location, away from all possible sources of ignition. Separate other chemicals particularly oxidizing gases, chlorure, bromine, iodine, and acids. Cylinders should stand in a well-ventilated location protected from direct sunlight and all possible percussion. Employees should be provided with impervious clothing, (gloves, aprons, boots, and face shields) and with chemical cartridge respirators.

Disposal and waste treatment — ''Put into large vessel containing water. Neutralize with HCl. Discharge into a sewer with sufficient water.''[2]

X. AMMONIUM NITRATE

CAS RN 6484522, CEE q: 5000 t.

Formula — NH_4NO_3.

Flammability — Moderate fire risk by spontaneous reaction. Powerful oxidizing agent. Self-ignition of mixtures of ammonium nitrate with easily oxidizable organic materials (acetic acid) or finely divided metals may take place at moderately elevated temperatures.

Explosion — Nitrates may explode when exposed to heat or flame or by spontaneous chemical reaction. It is less sensitive than other nitrates to impact and difficult to detonate. Ammonium nitrate has all the properties of other nitrates but it is also able to detonate by itself under certain conditions. Ammonium nitrate in combination with nitro compounds forms one of the major high explosives for military use.

Hazardous potentials — Heat and confinement may explode it; when heated to decomposition it emits highly toxic fumes of NO_x. It ignites by heating its mixtures with combustible materials, or in presence of zinc powder and water; it explodes by impacting its mixture with ammonium sulfate.

Incompatibilities — Powdered metals: P, S, NH_4NO_3, Al, Sb, Bi, Cd, Cr, Co, Cu, Fe, Pb, Mg, Mn, Ni, Sn, stainless steel, Zn, and alkali metals.

Toxicity — As for other nitrates, large amounts taken by mouth may have serious, or even fatal effects. Small, repeated doses may lead to weakness, general depression, headache, and metal impairment. Also there is some implication of increased cancer incidence among those exposed. It is an allergen.

Special precautions and preventive measures — Protect against physical damage. Store in well-ventilated buildings, preferably of noncombustible construction and equipped with automatic sprinkler protection. Floor drains and recesses should be plugged or eliminated to prevent entrapment of flowing molten nitrate during fire. Separate from all organic materials or other contaminating substances such as flammable liquid acids, corrosive liquids, organic chemicals, chlorates, sulfur, and finally divided metals or charcoal, coke, and sawdust. Open flames, sparks, and other sources of heat should be kept 50 ft away storage areas of bags. Wear rubber gloves.

Disposal and waste treatment — "Remove slowly into a large container of water. Add soda ash slightly by stirring. After 24 h, decant or siphon into another container. Neutralize with 6 M HCl and drain into the sewer with abundant water."[2]

XI. ANABASINE (3-(2-PIPERIDINYL) PYRIDINE, NEONICOTINE)

CAS RN, CEE q.
Formula — $C_{10}H_{14}N_2$.
Toxicity — It is used as an insecticide; it may cause acute and subacute toxicity, increased salivation, vertigo, confusion, disturbed vision and hearing, photophobia, cold extremities, nausea, vomiting, syncope, and chronic spasmus.

XII. ANTIMONY HYDRIDE (STIBINE)

CAS RN 7803523 CEE q: 100 Kg.
Formula — SbH_3.
Flammability — Moderate fire hazard when exposed to flame. It ignites on heating.
Explosion — Violent reaction with Cl_2, HNO_3, O_3, and NH_3.
Hazardous potentials — When heated to decomposition it emits hydrogen, metallic Sb, and toxic fumes of Sb. Hazardous reactions with hydrogen: explosion.
Toxicity — TLV Air: 0.1 ppm and OSHA standard air: TWA: 0.5 mg/mc = (0.1 ppm), averaged over an 8-h work shift. As other antimony compounds, stibine may affect the nervous system. Locally it is irritant to the skin and mucous membrane.
Special precautions and preventive measures — Provide an adequate ventilation. Wear rubber gloves, chemical goggles, chemical cartridge respirator, and coveralls.
Disposal and waste treatment — "Dissolve in a minimum amount of concentrated hydrochloric acid. Add to water until the appearance of white precipitate. Add 6 M HCl just to dissolve again. Saturate with hydrogen sulfide. After filtration, wash the precipitate, dry, package and return to suppliers."[2]

XIII. ARSENIC

CAS RN 7440382, CEE q: *****.
Formula — As.
Flammability — Moderate fire hazard, in the form of dust, when exposed to heat, flame, or by chemical reactions with powerful oxidizers.
Explosion — Slight hazard when exposed to flame (in the form of dust).

Hazardous potentials — When heated or on contact with acid or acid fumes it emits highly toxic fumes; it can react vigorously on contact with oxidizing materials.

Incompatibilities — Bromine oxide, dirubidium acetylide, halogens, palladium, zinc, platinum, NCl_3, Ag NO_3, GO_3, and Na_2O_2.

Toxicity — As with other metallic poisons, the toxicities especially the acute toxicities, of arsenic compounds are related to their solubility in water. Systemic arsenic poisoning is rarely seen in industry, and still more rarely is it severe in character. It is difficult to explain the difference between industrial and nonindustrial arsenic poisoning, but such variation is recorded in all industrialized countries. The usual effects on worker are local, on skin and mucous membranes; perforated nasal septum is a common result of prolonged inhalation of white dust or fume. It is a human carcinogenic substance. Arsenic is highly toxic via intramuscular, subcutanes, intraperitoneal, inhalation, and oral routes; it may cause dermatitis, effects on the central nervous, gastrointestinal, genitourinary, and hemopoietic systems. It is used as food additive in food for human ingestion. A search of the world literature reveals no reports of industrial or experimental exposures solely to arsenic compounds which contain both environmental and toxicological criteria from which a TLV can be unequivocally based. TLV-TWA: 200 $\mu g/m^3$; OSHA standard: air TWA: 500 $\mu g/m^3$; and occupational exposure to inorganic arsenic recommended standard: CL 2 $\mu g/m^3$.

Special precautions and preventive measures — Protect containers against physical damage. Store in well-ventilated area away from food or food products and combustible materials. The work is preferably done in closed systems. Wear mechanical filter respirator, rubber gloves, boots, long-sleeved coveralls, and protective cotton clothing laundered daily. Physical examinations of exposed personnel should be periodically available.

Disposal and waste treatment — ''Dissolve in a minimum amount of concentrated HCl. Add to water until the appearance of white precipite. Saturate with hydrogen sulfide. After filtration, wash the precipitate, dry, package, and return to suppliers.''[2]

XIV. ARSENIC PENTOXIDE (ARSENIC ANHYDRIDE)

CAS RN 1303282, CEE q: 500 kg.
Formula — As_2O_5.
Hazardous potentials — When heated to decomposition it emits toxic fumes of As. It reacts vigorously with Rb_2C_2.
Toxicity — Carcinogenic determination: human positive (IARC). OSHA Standard Air: TWA 50.0 μg; occupational exposure to inorganic arsenic recommended standard: air CL 2 $\mu g/m^3$/ISM. Arsenic pentoxide is highly toxic via oral and intravenous routes. It is a poison. Acute allergic reactions to arsenic compounds used in medical therapy have been fairly common.

XV. ARSENIC TRIOXIDE (ARSENIOUS ACID, ARSENIOUS ANHYDRIDE)

CAS RN 1327533, CEE q: 100 kg.
Formula — As_4O_6.
The Arsenic Trioxide is used in the manufacture of glass and in agriculture as an insecticide and an herbicide.
Hazardous potentials — Reacts vigorously with Rb_2C_2, ClF_3, F_2, Hg, OF_2, and $NaClO_3$.
Toxicity — Carcinogenic determination: human positive (IARC). TLV: air 250 $\mu g/m^3$ (as As); OSHA standard air: TWA 500 μg (As)/m^3; and occupational exposure to inorganic arsenic recommended standard: air: CL 2 $\mu g/m^3$/15 min. Arsenic trioxide is a poisoning substance as other arsenic compounds; for acute poisoning: see Section XVI. Chronic poisoning may attack skin, central nervous system, gastrointestinal system, genitourinary system, and hematopoietic systems.

TCL_0: 110 μg (As)/m³ (via inhalation route; skin effects).
LDL_0: 1 mg/kg (via oral route).
TCL_0: 200 μg (As)/m³ (via inhalation route).

XVI. ARSINE

CAS RN 7784421, CEE q: 10 kg.
Formula — AsH_3.
Flammability — Moderate fire hazard when exposed to flame.
Explosion — Moderate explosion hazard when exposed to CL_2, HNO_3, (K + NH_3), or to open flame.
Hazardous potentials — When heated to decomposition it emits highly toxic fumes.
Incompatibilities — Oxidizing materials.
Toxicity — The toxicity of arsine is due to its hemolytic action. On entering the bloodstream it combines with the hemoglobin of the red blood cells; gradually the arsenic in this hemoglobin-arsenic complex is oxidized and the oxidation process is accompanied by hemolysis of the cell. The resulting anemia is responsible for the production of many of the symptoms. Frequently there is edema of the lungs, which may be accompanied by cyanosis.
TDL_0: 338 ppt (administered to man via inhalation route; gastrointestinal effects; systemic effects; effects on the central nervous system).
TLV: 0.2 mg/m³ (0.05 ppm) and OSHA standard: air: TWA 0.05 ppm.
TCL_0: 3 ppm (via inhalation route).
TCL_0: 0.5 ppm (via inhalation route).
LCL_0: 25 ppm/30 min (via inhalation route).
Occupational exposure to inorganic arsenic recommended standard: air CL 2 μg/m³/15 min.
Special precautions and preventive measures — Employ wet methods where possible. Provide an adequate ventilation (with downward exhaust). Wear protective cotton clothing laundered daily, safety goggles, air line mask, and rubber gloves. Compulsory bathing at the end of the work shift.
Disposal and waste treatment — "In case of leakage, dilute with inert gas and put into a hood for ventilation. Dispose by controlled burning or, for little quantities, seal cylinders and return to suppliers."[2]

XVII. AZINPHOS METHYL

CAS RN: 86500, CEE q: 100 kg.
Formula — $C_{10}H_{12}N_3O_3PS_2$.
Hazardous potentials — When heated to decomposition it emits toxic fumes and vapors (such as sulur dioxide, oxides of nitrogen, phosphoric acid mist, and carbon monoxide). Contact with strong oxidizers may cause fires and explosions.
Toxicity — Azinphos-methyl is an anticholinesterase agent; absorption may occur from inhalation of the dust or mist, from skin absorption of solutions, or from ingestion. Signs and symptoms of overexposure are caused by the inactivation of the enzyme cholinesterase which results in the accumulation of acetylcholine at synapses in the nervous system, skeletal and smooth muscles, and secretary glands. It is a food additive permitted in the food and drinking water of animals and/or for the treatment of food-producing animals. Acceptable daily intake: 0.0025 mg/kg body weight.

Commodity	MRL (mg/kg)
Alfalfa (green)	2
Almond hulls	10

Commodity	MRL (mg/kg)
Almond	0.2 (on a shell-free basis)
Apricots	2
Broccoli	1
Brussel sprouts	1
Celery	2
Cereal grains	0.2
Citrus fruit	2
Cottonseed	0.2
Fruit (except as otherwise listed)	1
Grapes	4
Kiwi fruit	4 (whole fruit) 0.4 (edible part)
Melons	2
Peaches	4
Pea vines	2
Potatoes	0.2
Soybeans (dry)	0.2
soybean vines	2
Sunflower seed	0.2
Vegetables	0.5

TLV-TWA: 0.2 mg/m^3; TLV-STEL: 0.6 mg/m^3; and OSHA standard air TWA: 0.2 mg/m^3 averaged over an 8-h work shift.

Precautions and preventive measures — Impervious clothing, gloves, face shields and other appropriate protective clothing necessary to prevent skin contact with azinphos methyl or liquids containing azinphos methyl.

Disposal and waste treatment — ''Azinphos methyl may be disposed: (1) by making packages of azinphos methyl in paper or other flammable material and burning in a suitable combustion chamber or (2) by dissolving azinphos-methyl in a flammable solvent and atomizing in a suitable combustion chamber.''[3]

XVIII. AZOCICLOTIN (1H-1,2,4-TRIAZOLYL) 1-1 TRICYCLO HEXYLSTANNANE

CAS RN 41085118, CEE q: 100 kg.
Formula — $C_{20}H_{35}N_3Sn$.
Hazardous potentials — When heated to decomposition it emits toxic fumes of NO_x.
Toxicity — OSHA standard 0.1 mg/mc. It is employed as a pesticide.
FAO maximum residue limits are

Commodity	MRL (mg/kg)
Apple	2
Grapes	2
Strawberry	2
Bean	0.5

XIX. BARIUM AZIDE

CAS RN 18810587, CEE q: 50 t.
Formula — BaN_6.

Flammability — Spontaneously flammable in air at around 275°C.
Explosion — Moderate explosion hazard when shocked or exposed to heat. Very unstable.
Hazardous potentials — Dangerous.
Toxicity — Aquatic toxicity rating TLm_{36} 100.10 ppm.
Special precautions and preventive measures — Do not store on road with any high explosive. (azides).

XX. BENZENAMINE (ANILINE)

CAS RN 62533. CEE q.
Formula — C_6H_7H.
Flammability — Flash point 76°C (closed cup), ignition temperature 700°C, and flammable limit (lower): 1.3%. Combustible.
Explosion — Vapor forms explosive mixtures with air.
Hazardous potentials — It may ignite by violent reaction with concentrated nitric acid or fuming nitric acid. When heated to decomposition it emits highly toxic fumes of NO_x, and CO.
Incompatibilities — Contact of liquid aniline with strong acids will cause violent spattering. Contact with strong oxidizers may cause fires and explosions.
Toxicity — It is highly toxic when absorbed through skin, accidentally inhaled, or swallowed. Its most important action is the formation of methemoglobin, with the resulting anoxemia and depression of the central nervous system. TLV:TWA 2 ppm (skin), OSHA standard TWA = 19 mg/mc, and LDL_0: 357 mg/kg.
Special precautions and preventive measures — Liquid aniline will attack some forms of plastics, rubber, and coating. Operations should be enclosed as much as possible. Store in cool, dry, well-ventilated locations, away from possible contact with oxidizing materials. Use chemical cartridge or airline respirator. Wear, when it is necessary, protective clothing, laundered daily in addition to synthetic butyl boots, gloves, safety goggles, and hat.
Disposal and waste treatment — (1) "Dissolve in such combustible solvent as alcohols, benzene etc.; spray the solution into the furnace with after burner and scrubber, or (2) Put into a mixture of sand and soda ash. After mixing put into a paper carton with paper. Burn in a furnace."[2] "Absorb in a vermiculite, dry sand, earth, or similar material and dispose in a secured sanitary landfill."[3]

XXI. BENZIDINE

CAS RN 92875, CEE q: 1 kg.
Formula — $C_{12}H_{12}N_2$.
Hazardous potentials — When heated to decomposition it emits highly toxic fumes of NO_x.
Toxicity — Highly toxic via oral and probably inhalation and subcutaneous routes. It may cause damage to blood, including hemolysis and bone marrow depression. It is a carcinogen substance. Carcinogenic determination: human positive (IARC). OSHA: carcinogen, and TLV: any exposure is considered extremely hazardous.
Special precautions and preventive measures — Provide an adequate ventilation. Wear chemical goggles, mechanical filter respirator, rubber gloves, and other adequate protective clothing laundered daily with compulsory bathing at the end of the work shift.
Disposal and waste treatment — "(1) Dissolve in such a combustible solvent as alcohols, benzene, spray the solution into the furnace. (2) Pour into a mixture of sand and soda ash; put into a paper carton with other paper and burn in a furnace."[2]

XXII. BERYLLIUM

CAS RN 7440417, CEE q: 10 kg.

Formula — Be.

Flammability — Moderate fire hazard when beryllium is in form of dust or powder in air, or when it is exposed to flame, or by spontaneous chemical reactions.

Explosion — Dust may produce an explosive mixture with air.

Hazardous Potentials — It reacts with strong acids to evolve hydrogen. When heated to decomposition in air it emits very toxic fumes of BeO.

Toxicity — TLV-TWA: 0.002 mg/m^3. OSHA standard air TWA 2 μg/m^3, Cl 5 μg/m^3, and PIC: 25 μg/m^3 30 min/8-h work shift. Occupation exposure to Be: recommended standard and air: Cl 0.5 μg/m^3/130 min.

TCL$_0$: 300 mg/m^3 (pulmonary effects).

TCL$_0$: 0.1 mg/m^3 (pulmonary effects).

Special precautions and preventive measures — Store in well-marked area to avoid beryllium or beryllium compounds being mistaken for other materials. Care should be taken to avoid an excessively moist atmosphere which could accelerate formation of oxide. Protect containers against physical damage. Keep away from acids, chlorinate of hydrocarbons, and oxidizable materials. Wear chemical goggles, mechanical filter, respirator, and rubber gloves. Provide adequate ventilation, strict measures suppressing dust. Preclude from exposure those individuals with previous pulmonary diseases. Examination for exposed personnel should be made available.

Disposal and waste treatment — "Dissolve in a minimum amount of 6 *M* HCl and filter. Add the filtrate to slight excess of 6*M* NH$_4$OH by use of the litmus test. Heat and coagulate the precipitate. After 12 h filter and dry. Package and return to the suppliers."[2]

XXIII. BIS-(2-CHLOROETHYL) SULFIDE

CAS RN 505602, CEE q: 1 kg.

Formula — C$_4$H$_8$ Cl$_2$S.

Flammability — Moderate fire hazard when exposed to heat or flame. It can be ignited by a large explosive charge.

Hazardous potentials — Dangerous; when heated to decomposition or on contact with acid or acid fumes it emits highly toxic fumes of oxides of sulfur and chlorides; it will react with water or steam to produce toxic and corrosive fumes; it can react vigorously with oxidizing materials.

Toxicity — Carcinogenic determination: human suspected IARC. In animals it is a highly toxic substance via inhalation, dermal, subcutaneous, intravenous routes. It is a highly irritant to the eye, skin, and lungs. Pulmonary lesions are often fatal. There can also be severe gastric disturbances. It damages blood vessels penetrating deeply. It produces marked leukopenia with involution of the lymphatic nodes.

LC$_{50}$: 1500 mg/m^3/min (via inhalation route).

LCL$_0$: 23 ppm/10 min (via inhalation inhalation).

LDL$_0$: 64 mg/kg (by skin).

XXIV. BIS(CHLOROMETHYL)ETHER

CAS RN 542881, CEE q: 1 kg.

Formula — C$_2$H$_4$Cl$_2$O.

Flammability — Flash point below 19°C.

Hazardous potentials — When heated to decomposition it emits very toxic fumes of Cl.

Toxicity — It is highly toxic via inhalation route. bis(chloromethyl)ether is a recognized human carcinogen with an assigned TLV. TLV-TWA: 0.001 ppm (in air). Several other nations class it as a carcinogen with no limit.

Special precautions and preventive measures — Provide an adequate ventilation. Store in well-ventilated place away from source of ignition. Wear chemical goggles, chemical respirator, rubber gloves, and other protective clothes.

XXV. BIS(DIMETHYLAMIDO)FLUOROPHOSPHATE

CAS RN 115264, CEE q: 100 kg.

Formula — $C_4H_{12}FN_2OP$.

Hazardous potentials — When heated to decomposition it emits very toxic fumes of F, NO_x, PO_x.

Toxicity — Highly toxic via oral, dermal, intraperitoneal, subcutaneous, or intravenous routes. It is a cholinesterase inhibitor.

LD_{50}: 1 mg/kg (administered to rat via oral route).

LD_{50}: 2 mg/kg (administered to rat by skin).

LD_{50}: 5 mg/kg (administrated to rat via intraperitoneal route).

LDL_0: 300 µg/kg (administered to rat via subcutaneous route).

LD_{50}: 2 mg/kg (administered to mouse via oral route).

LD_{50}: 1400 µg/kg (administered to mouse intraperitoneal route).

LD_{50}: 1 mg/kg (administered to mouse via subcutaneous route).

LD_{50}: 5 mg/kg (administered to dog via intravenous route).

LD_{50}: 2 mg/kg (administered to cat via oral route).

LD_{50}: 3 mg/kg (administered to rabbit via oral route).

LD_{50}: 6 mg/kg (administered to rabbit via subcutaneous route).

LD_{50}: 3 mg/kg (administered to rabbit via intravenous route).

LD_{50}: 4 mg/kg (administered to guinea pig via oral route).

LD_{50}: 2500 µg/kg (administered to guinea pig via intraperitoneal route).

LD_{50}: 2 mg/kg (administered to guinea pig via subcutaneous route).

XXVI. BROMINE

CAS RN 7726956, CEE q: 500 t.

Formula — Br_2.

Flammability — Nonflammable.

Hazardous potentials — It is a strong oxidizing material; it reacts spontaneously with reducing agents and the heat of reaction may cause ignition of combustible materials. When heated it emits highly toxic fumes. It will react with water and steam to produce toxic and corrosive fumes.

Incompatibilities — Contact of bromine with aqueous ammonia may cause violent reactions. Bromine will react violently (ignition and explosion) with aluminum, titanium, mercury, potassium, hydrogen, methane, ethylene, sulfur, antimony, arsenic, phosphor, and sodium. It may cause explosion by contact with combustible material and metallic powder.

Toxicity — Bromine is highly toxic if it is inhaled or if it comes in contact with the eyes or skin. Vapor is irritating to the eyes and respiratory tract. Pulmonary edema and pneumonia may be a delayed complication of severe exposures. In humans, 10 ppm is intolerable, causing severe irritation of the upper respiratory tract. TLV-TWA Air 0.1 ppm (0.7 mg/mc); TLV-STEL: 0.5 ppm (2 mg/m³); and OSHA standard air TWA 0.1 ppm.

LDL_0 14 mg/kg (via oral route).

LCL_0 100 ppm (via inhalation route).

Special precautions and preventive measures — Protect containers against physical damage. Store in cool and dry areas out of direct sunlight. Separate from combustible, organic, or other readily oxidizable materials. Keep above $-6.7°C$ to prevent freezing but avoid heating above atmospheric temperatures as vapor pressure increase could rupture the containers. Wear rubber gloves, face shield, chemical goggles, coveralls, self-contained breathing apparatus, and boots. Regular physical examinations should be made upon people who work with bromine or bromides.

Disposal and waste treatment — "Use vast volume of concentrated solution of reducing agent (bisulfites or ferrous salts with 3 M H_2SO_4 or hypo). Neutralize with soda ash or dilute HCl. Drain into the sewer abundant water."[2]

XXVII. BROMOMETHANE (METHYL BROMIDE)

CAS RN 74839, CEE q:200 t.

Formula — CH_3Br.

Flammability — It has fire and flammable risk. Flash point: none in standard test method. Autoignition temperature: 537°C. Flammable limits: (percent by volume in air) 13.5 to 14.5%, only in presence of high energy ignition source.

Explosion — Moderate explosion risk when exposed to sparks or flame.

Hazardous potentials — When heated to decomposition it emits highly toxic fumes of hydrogen bromide and carbon monoxide.

Incompatibilities — Contact with aluminum and strong oxidizers may cause fires and explosions. Form explosive mixtures with air within narrow limits at atmospheric pressure, but wider at higher pressure.

Toxicity — Methyl bromide gas is a severe pulmonary irritant and neurotoxic; it is a powerful fumigant gas, one of the most toxic of the common organic halides. It is hemotoxic and narcotic with delayed action; it is cumulative and damaging to nervous system, kidneys, and lung. Central nervous system effects include blurred vision, mental confusion, numbness, tremors, and speed defects. Methyl bromide is reported to be eight times more toxic on inhalation than ethyl bromide. Moreover, because of its greater volability, methyl bromide is a much more frequent cause of poisoning; pulmonary edema is the chief problem in acute poisoning. Repeated splashes of the liquid on the skin cause marked irritation and vesciculation.

LCL_0: 60,000 ppm/2 h (via inhalation route).

LCL_0: 5 mg/m³/2 h (administered to child via inhalation route).

TCL_0: 35 ppm (via inhalation route; gastrointestinal effects).

TLV-TWA: 5 ppm (skin) and OSHA standard air: CL 20 ppm (skin).

Physical examination of exposed personnel every 6 months with special attention to central nervous system should be available.

Special precautions and preventive measures — Store in cool place (T < 40°), kept away from direct sunlight and avoid all the heat sources. Liquid methyl bromide will attack some forms of plastics, rubber, and coatings. It flames up by an impact (with alkali metals). Provide an adequate ventilation. Wear chemical goggles, gloves, chemical cartridge, or airline respirator.

Disposal and waste treatment — "Dissolve in a combustible solvent. Scatter the spray of the solution into the furnace with afterburner and alkali scrubber."[2]

XXVIII. *n*-BUTYL ACETATE

CAS RN 123864, CEE q: 200 t.

Formula — $CH_3COO(CH_2)_3CH_3$.

Flammability — Dangerous fire and explosion risk. Flash point: 23°C (closed cup), 38°C (open cup); ignition temperature: 421°C; and flammable limits: 1.7 to 15%. The ester does not undergo reaction on fire because of its chemical stability.

Explosion — Contact with nitrates, strong oxidizers, strong alkalies, and strong acids may cause fires and explosions. When heated to decomposition it emits acid smoke and irritant fumes such as carbon monoxide.

Toxicity — Butyl acetate can affect the body via inhalation dermal routes, or if it comes in contact with the eyes or skin. The principal effect of over-exposure to butyl acetate is irritation of the eyes and nose, which occurs at 200 to 300 ppm and is marked at concentration over 3000 ppm. It may cause narcosis. TLV-TWA: air 150 ppm; TLV-STEL: 200 ppm; and OSHA standard air: TWA 150 ppm averaged over an 8-h work shift.

TCL_0: 200 ppm (administrated via inhalation route, irritant effects).

TLV: sec-Butyl Acetate) 200 ppm

Preventive measures — Provide an adequate ventilation. Store in cool well-ventilated place, away from sources of ignition. Separate from explosive oxidizing materials, organic peroxides, and foodstuffs. Wear chemical safety glasses, chemical cartridge mask, and other rubber protective clothing.

Disposal and waste treatment — "Spray into the furnace; incineration will become easier by mixing with a more flammable solvent."[2]

XXIX. *t*-BUTYL PEROXYACETATE (*t*-BUTYL PERACETATE)

CAS RN 107711, CEE q: 50 t.

Formula — $C_6H_{12}O_3$.

Flammability — It is dangerous on contact with heat, flame, reducers. Flash point: < 80°F ≈ 26.7°C.

Explosion — Pure ester, it is shock sensitive and it detonates. It explodes with great violence when rapidly heated to critical temperature. *t*-Butyl peracetate can explode in contact with organic matter.

Hazardous potentials — It is sensitive to shock and heat.

Toxicity — Moderately toxic via oral and inhalation routes; as other peroxides it may cause injury on contact with skin or mucous membranes.

LD_{50}: 632 mg/kg (administered to mouse via oral route.

LCL_0: 6000 mg/m^3 (administered to mouse via inhalation route).

XXX. *TERT*-BUTYL-PEROXYPIVALATE

CAS RN 927071, CEE q: 50 t.

Formula — $C_9H_{18}O_3$.

Flammability — Moderate fire hazard via heat, flame, or sparks. Dangerous by chemical reactions with reducing agents or exposure to heat. It is as other peroxides, a powerful oxidizer.

Explosion — It causes a severe explosion hazard when shocked, when exposed to heat, or by spontaneous chemical reaction, upon contact with reducing materials, such as organic matter or thiocyanates.

Toxicity — As other peroxides it is irritant to the skin, eyes and mucous membrane.

LD_{50}: 4300 mg/kg (administered to rat via oral route).

Special precautions and preventive measures — Store in cool, well-ventilated locations. Wear gloves, chemical goggles, face shields, and respirator.

XXXI. CARBOFENOTHION (TRITHION)

CAS RN 786196, CEE q: 100 kg.
Formula — $C_{11}H_{16}ClO_2PS_3$.
Flammability — Vapor pressure is extremely low.
Hazardous potentials — When heated to decomposition it emits very toxic fumes of SO_x, PO_x.
Toxicity — Carbofenothion is highly toxic via oral route and by skin contact. It acts as a cholinesterase inhibitor. It is used as insecticide and acaricide. It is a food additive permitted in feed and drinking water of animals and/or for the treatment of food-producing animals. Acceptable daily intake: 0.0005 mg/kg body weight. Maximum residue limits (definition of residue: sum of carbophenothion, its sulfoxide and its sulfore expressed as carbophenothion):

Commodity	MRL (mg/kg)
Apples	1
Apricots	1
Broccoli	0.5
Brussel sprouts	0.5
Cattle, carcass meat (in the carcass fat)	1
Cauliflower	0.5
Citrus fruit	2
Milk	0.004 F
Nectarines	1
Olive oil	0.2
Olives (unprocessed)	0.1
Peaches	1
Pears	1
Pecans (on a shell-free basis)	0.02
Potatoes	0.02
Plums	1
Rapeseed	0.02
Sheep, carcass meat (in the carcass fat)	1
Spinach	2
Sugar beets	0.1
Walnuts (on a shell-free basis)	0.02

Special precautions and preventive measures — Provide adequate ventilation, wear self-contained breathing apparatus and overalls protective clothing, chemical goggles, rubber gloves, and shoes.

XXXII. CARBOFURAN

CAS RN 1563662, CEE q: 100 kg.
Formula — $C_{12}H_{15}NO_3$.
Hazardous potentials — When heated to decomposition it emits highly toxic fumes of NO_x.
Toxicity — It is highly toxic via oral route. It may be absorbed through skin. It is a carbamate insecticide. It acts as a cholinesterase inhibitor. Acceptable daily intake: 0.01 mg/kg body weight. Maximum residue limits (sum of carbofuran and 3-hydroxycarbofuran expressed as carbofuran):

Commodity	MRL (mg/kg)
Alfalfa (fresh)	5
Alfalfa hay	20
Banana	0.1
Barley	0.1
Brussel sprout	2
Cabbage	0.5
Carrot	0.5
Cattle, carcass meat	0.05
Cattle, fat	0.05
Cattle, meat by-products	0.05
Cauliflower	0.2
Coffee (raw beans)	0.1
Eggplant (aubergines)	0.1
Goat, fat	0.05
Goat, meat by-products	0.05
Goat, carcass meat	0.05
Hops (dried)	5
Horse, carcass meat	0.05
Horse, fat	0.05
Horse, meat by-products	0.05
Kohlrabi	0.1
Lettuce	0.1
Maize fodder (fresh)	5
Maize	0.1
Milk	0.05
Mustard seed	0.1
Oats	0.1
Oilseed	0.1
Onion	0.1
Peach	0.1
Peanut (kernels)	0.1
Pear	0.1
Pig, carcass meat	0.05
Pig, fat	0.05
Pig, meat by-products	0.05
Potato	0.5
Rice (hulled)	0.2
Sheep, carcass meat	0.05
Sheep, fat	0.05
Sheep, meat by-products	0.05
Sorghum	0.1
Soybean	0.2
Strawberry	0.1
Sugar beet leaves	0.2
Sugar beet (roots)	0.1
Sugar cane	0.1
Sweet corn (kernels)	0.1
Tomato	0.1
Wheat	0.1

TLV-TWA 0.1 mg/mc.

DL_{50}: 11 mg/kg (via oral route).

DL_{50}: 10 mg/kg (by skin).

Preventive measures — Wear chemical goggles, chemical cartridge respirator, and pro-tective clothes. Use adequate ventilation.

XXXIII. CARBON DISULFIDE

CAS RN 75150, CEE q: 200 t.
Formula — CS_2.
Flammability — Very flammable liquid, dangerous when exposed to heat, flame, sparks, or friction; flash point: $-30°C$ (closed cup); ignition temperature: $100°C$; and flammable limits: 1 to 50% by volume in air. It may travel a considerable distance to a source of ignition and flash back. Vapors may be ignited by contact with an ordinary light bulb or by friction.

Explosion — Severe explosion hazard when exposed to heat or flame; it reacts violently (fires, explosions) with Al, Cl_2, F_2, Na, R, and Zn.

Hazardous potentials — When heated to decomposition it emits highly toxic fumes of SO_x; it can react vigorously with oxidizing materials. Imcompatibilities: air halogens.

Toxicity — TLV-TWA: 10 ppm (skin). OSHA Standard: air Twa 20 ppm averaged over an 8-h work shift. CL = 30 ppm and K = 100 ppm/30 min in an 8-h work shift. Occupational exposure to carbon disulfide, recommended standard: Air TWA: 2 ppm and Cl: 10 ppm/15 min. Carbon disulfide vapor is absorbed largely through the lungs and through skin. It is employed as an insecticide. The chief toxic effect is on the central nervous system, acting as a narcotic and anesthetic in acute poisoning with death following from respiratory failure. The anesthetic action is much more powerful than that of chloroform. In acute poisoning it may cause a vesicant action skin. Carbon disulfide intoxication can involve all parts of the central and peripheral nervous system including damage to cranial nerves and development of polyneuritis with paresthesias and muscles weakness in the extremities, unsteady gait, and dysphagia. A Parkinson-like syndrome can result. Psychosis and suicide are established risks of overexposure to carbon disulfide. The lowest concentration causing demonstrated health effect (cardiovascular, neurologic) is 10 ppm. Vision, gastrointestinal system, and genitourinary system, may also be attempted.

Special precautions and preventive measures — Liquid carbon disulfide will attack some forms of plastics rubber and coatings. Protect containers against physical damage. Store in well-detached and isolated places from other buildings, other materials, and possible sources of ignition, preferably in a building of noncombustible material or better, in a construction with floor level ventilation. Avoid direct sunlight. Provide an adequate ventilation. Tanks should be submerged in water or located over concrete basins containing water or sufficient capacity to hold all of the tank contents in addition to the water. Water or inert gas should be provided over the carbon disulfide in all tanks. No electrical installation or heating facilities should be permitted in or near storage area. Protect against lightning and static electricity. Wear chemical goggles, airline or chemical cartridge respirator, rubber gloves, aprons and boots, and face shields.

Disposal and waste treatment — "(1) Pay special attention to avoid ignition by static charge. Absorb in sand or ash, and cover with water. Transfer the contents under water in buckets to an open area. Burn with care by igniting from a safe distance or (2) recover by distillation."[2]

XXXIV. CHLORINE

CAS RN 7782505, CEE q: 50 t.
Formula — Cl_2.
Flammability — Chlorine is not combustible in air, but most combustible materials will burn in chlorine; flammable gases and vapors will form explosive mixtures with chlorine.

Explosion — It reacts explosively, or forms explosive compounds with many common chemicals, especially acetylene, turpentine, ether, ammonia gas, fuel gas, hydrocarbon, hydrogen, and finely pulverized metals except the noble gases and carbon (in the absence of combined hydrogen), and also with Al, Sb, As, Br, and B.

Hazardous potentials — When heated it emits highly toxic fumes; it will react with water or steam to produce toxic and corrosive fumes of hydrogen chloride.

Toxicity — Chlorine can affect the body if it is inhaled or if it comes in contact with the eyes or skin. It is highly irritant to the mucous membrane of the eyes and to the respiratory tract. Severe exposure may cause pneumonia and may also be fatal. Repeated or prolonged exposure to chlorine may cause corrosion of the teeth, skin irritation, difficult breathing, bronchitis, and other chronic lung conditions or irritations of the upper respiratory tract. Preclude from exposure those individuals with pulmonary diseases. Periodical physical examinations including chest X-ray, should be made available. TLV-TWA: 1 ppm; TLV-STEL: 3 ppm; level of odor perception: 3.5 ppm; OSHA standard air TWA: 1 ppm, averaged over a 15-min period; and occupational exposure to chlorine recommended standard: air: Cl 0.5 ppm/15 min;

TCL_0: 15 ppm (administered via inhalation route).

LCL_0: 430 ppm/30 min (administered via inhalation route).

LCL_0: 873 ppm/30 min (administered via inhalation route).

Special precautions and preventive measures — Extremely strong oxidizing agent; it will attack some forms of plastics, rubber, and coating. Protect containers against physical damage. Cylinders and tin containers should be stored in cool, dry, relatively isolated areas; protect from weather and extreme temperature changes. Protect containers from external heat. Separate from combustible, organic, or easily oxidizable materials and especially isolate from acetylene, ammonia hydrogen, hydrocarbons, ether turpentine, and finely divided metals. Store in outdoor or in well-ventilated noncombustible construction, detached or segregated. Provide adequate ventilation. Wear chemical goggles and other adequate impervious protective clothing; chemical cartridge respirator.

Disposal and waste treatment — Use vast volume of concentrated solution of reducing agent. Neutralize with soda ash or dilute HCl. Drain into the sewer with abundant water. Allow gas to dispers at a safe location.

XXXV. CHLOROMETHYL METHYL ETHER

CAS RN 107302, CEE q: 1 kg.

Formula — C_2H_5ClO.

Flammability — Flammable liquid. Flash point: <73.4°F ≈ 23°C.

Hazardous potentials — When heated to decomposition it emits toxic fumes of Cl.

Toxicity — Aquatic toxicity rating TLm_{96}: 1000 to 100 ppm. Carcinogeric determinations: human suspected (IARC). Carcinogen (OSHA). Highly toxic via inhalation route in rat and hamster; moderately toxic in rat via oral route. It acts as a DNA inhibitor.

LD_{50}: 817 mg/mc (administered to rat via oral route).

LC_{50}: 55 ppm (administered to rat via inhalation route over a period of 7 h).

LC_{50}: 65 ppm (administered to hamster via inhalation route over a period of 7 h).

Special precautions and preventive measures — Wear rubber gloves, face shield, and respirator. Separate from any source of ignition.

XXXVI. CHLOROPHENVINPHOS (DIETHYL 1-(2,4 DICHLOROPHENYL)-2-CHLOROVINYL PHOSPHATE)

CAS RN 470906, CEE q: 100 kg.

Formula — $C_{12}H_{14}Cl_3O_4P$.

Hazardous potentials — When heated to decomposition it emits very toxic fumes of Cl and PO_x.

Toxicity — A cholinesterase inhibitor. Highly toxic via oral, dermal, intraperitonal, subcutaneous, or intravenous routes. See also Section CXII. Acceptable daily intake: 0.002 mg/kg body weight. Maximum residues limits (definition of residue: chlorfenvinphos, sum of E- and r-isomers. Fat soluble residues):

Commodity	MRL (mg/kg)
Broccoli	0.05
Brussel sprouts	0.05
Cabbage	0.05
Carcass meat (in the carcass fat)	0.2
Carrots	0.4
Cauliflower	0.1
Celery	0.4
Cottonseed	0.05
Eggplants (aubergines)	0.05
Horseradish	0.1
Leeks	0.05
Maize (in the kernels)	0.05
Milk	0.008
Mushrooms	0.05
Onions	0.05
Peanuts (on a shell-free basis)	0.05
Potatoes	0.05
Radishes	0.1
Rice	0.05
Swedes (rutabagas)	0.05
Sweet potatoes	0.05
Tomatoes	0.1
Turnips	0.05
Wheat	0.05

TDL_0: 10 mg/kg (administered to man by skin; blood effects).

Special precautions and preventive measures — Chlorphenvinphos is corrosive for metals. Provide an adequate ventilation; wear chemical goggles, respirator, gloves, and rubber boots.

XXXVII. COBALT

CASRN:7440484, CEE q:100 kg.

Formula — Co.

Flammability — Powder is flammable. Autoignition temperature: expecially prepared (the form prepared by reducing the oxides in hydrogen) very fine cobalt dust will catch fire at room temperature in air. Pyroforie cobalt reacts violently with acetylene, air, and NH_4NO_3.

Hazardous potentials — Contact of dust with strong oxidizers may cause fire and explosion.

Toxicity — Cobalt metal fume and dust can affect the body if they are inhaled or if they come in contact with the eyes or skin. It is also toxic via oral route. exposure to cobalt may cause allergic skin rash when inhaled; cobalt metal and dust may cause upper respiratory tract, irritation, and chronic interstitial pneumonitis. Cobalt is a chemical necessary to human life but dangerous at high concentrations. It is an experimented carcinogeric substance. TLV-TWA:0.05 mg/mc; TLV-STEL: 0.1 mg/m^3; and OSHA:TWA: 0.1 mg/m^3 averaged over an 8-h work shift.

Special precautions and preventive measures — Use mechanical filter respirator and adequate local exhaust ventilation systems. Employees should be provided with impervious clothing, gloves, face shields, and other appropriate protective clothing necessary to prevent repeated or prolonged skin contact. Initial and periodic medical examination should be made available to each employee who is exposed to cobalt metal fume and dust.

Disposal and waste treatment — Sort, classify, and put in a box properly labeled. Dispose in a secure sanitary landfill.

XXXVIII. CRIMIDINA (CASTRIX)

CAS RN 535897, CEE q: 100 kg.
Formula — $C_7H_{10}ClN_3$.
Hazardous potentials — When heated to decomposition it emits very toxic fumes of Cl and NO_x.
Toxicity — It can cause damage to the central nervous System and convulsions. Highly toxic via oral routes.
LDL_0: 5 mg/kg (via oral route).
LD_{50}: 1250 μg/kg (administered to rat via oral route).
LD_{50}: 5 mg/kg (administered to rabbit via oral route)
LD_{50}: 1200 μg/kg (administered to mouse via oral route).
LD_{50}: 2660 μg/kg (administered to guinea pig via oral route).
LD_{50}: 22500 μg/kg (administered to chicken via oral route).
Special precautions and preventive measures — Store in isolated areas. Wear protective clothes.

XXXIX. 2-CYANO-2-PROPANOL

CAS RN 75865, CEE q: 200 t.
Formula — C_4H_7NO.
Flammability — Severe fire risk if exposed to heat or flame.
Hazardous Potentials — Severe explosion risk on contact with sulfuric acid (H_2SO_4). Vapor forms explosive mixtures with air.
Toxicity — URSS (1970): 0.9 mg/mc (skin).
Special precautions and preventive measures — Wear respirator, gloves, and other protective clothes. Provide adequate ventilation. Store in well-detached isolated places from other buildings, and possible source of ignition.

XL. CYCLOTETRAMETHYLENETETRANITRAMINE

CAS RN 2691410, CEE q: 50 t.
Formula — $C_4H_8N_8O_8$.
Hazardous potentials — When heated to decomposition it emits toxic fumes of NO_x.
Toxicity — Highly toxic substance via oral and intravenous routes. If ingested it may cause moderate erythema and moderate edema.
LD_{50}: 1500 mg/kg (administered to mouse via oral route).
LDL_0: 40 mg/kg (administered to dog via intravenous route).
LD_{50}: 300 mg/kg (administered to guinea pig via oral route).
LD_{50}: 28 mg/kg (administered to guinea pig via intravenous route).

XLI. CYCLOTRIMETHYLENENITRAMINE (SYM-TRIMETHYLENETRINITRAMINE)

CAS RN 121824, CEE q: 50 t.

Formula — $C_3H_6N_6O_6$.

Flammmability — As other nitrates moderate fire hazard by spontaneous chemical reaction, it is a powerful oxidizing agent.

Explosion — Class A explosive. It is one of the most powerful high explosives in use today. It has more shattering power than TNT and is often mixed with TNT as a bursting charge for aerial bombs, mines, and torpedoes. Because if is easily initiated by mercury fulminate, it may be used as a booster.

Hazardous potentials — On decomposition it emits toxic fumes of NO_x. As other nitrates it is a powerful oxidizing agent which can react violently with reducing materials.

Toxicity — Oral administration may cause moderate erythema and moderate edema. Cases of epileptiform convulsions have been reported from exposure. TLV: air: 1.5 mg/m^3 and OSHA standard air TWA: 1500 μg/m^3.

XLII. DEMETON (DEMETON-O + DEMETON-S)

CAS RN 8065483, CEE q: 100 kg.

Formula — $(C_2H_5O)_2PSOC_2H_4SC_2H_5$.

Flammability — Data not available.

Hazardous potentials — Contact with strong oxidizers may cause fires and explosions. Toxic gases and vapors (such as sulfur dioxide, phosphoric acid mist, and carbon monoxide) may be released when demeton decomposes.

Toxicity — Demeton is a highly toxic insecticide; absorption may occur from inhalation of the vapor, by skin absorption, or by ingestion of the liquid. As for other organic phosphorus poisons, the actions of this compound and its metabolites are believed to be based upon the inhibition of the enzyme cholinesterase, this allowing the accumulation of large amounts of acetylcholine. The most serious consequence is paralysis of the respiratory muscles. The joint FAO/WHO Alimentarius Commission, at its 16th session, decided to withdraw the codex maximum residue limits for demeton. The MRL was converted into a guideline level. Little is known regarding the acute and chronic dosages which would be dangerous to man. Doses of organic phosphorus insecticide tend to be cumulative in their effects. TLV-TWA: Air 0.01 ppm (skin) and OSHA standard air: 0.1 mg/m^3 of air averaged over an 8-h work shift.

LDL$_0$: 240 μg/kg (via oral route).

Special precautions and preventive measures — Demeton will attack some forms of plastics, rubber, and coatings. Provide an adequate ventilation. Wear impervious clothing, gloves, face shield, and other appropriate protective clothes. Medical procedures should be made available to each employee who is exposed to demeton at potentially hazardous levels, periodically. Examine the cholinesterase weekly.

Disposal and waste treatment — (1) Place the disposal in an open furnace. Add equal parts of sand and crushed limestone, then cover with a combustible solvent. Ignite from a safe distance and burn with the utmost care, or (2) Place the disposal with the sand and crushed limestone into a paper carton. Burn in the furnace with afterburner and alkaline scrubber, or (3) absorb in vermiculite, dry sand, earth, or a similar material and dispose in a secured sanitary landfill, or (4) atomize in a suitable combustion chamber.

XLIII. DI-*SEC* BUTYL PEROXYDICARBONATE (*SEC*-BUTYL PEROXIDICARBONATE)

CAS RN 19910657, CEE q: 50 t.
Formula — $C_{10}H_{18}O_6$.
Flammability — As other peroxides, *sec*-butyl peroxidicarbonate is dangerous by chemical reaction with reducing agent or exposure to heat; it is a powerful oxidizer.
Explosion — Severe explosion hazard when shocked, exposed to heat, or by spontaneous chemical reactions as for other organic peroxides; it may explode on contact with reducing materials.
Hazardous potentials — When heated to decomposition it emits acrid smokes and irritant fumes.
Toxicity — As other organic peroxides it is irritant to the skin, eyes, and mucous membrane.
LD_{50}: 1200 mg/kg (administered to rabbit by skin).

XLIV. 2,2-DI(*TERT*-BUTYLPEROXY)BUTANE

CAS RN 2167239, CEE q: 50 t.
Formula — $C_{12}H_{26}O_4$.
Hazardous potentials — Pure material explodes on heating, sparking, or impact.

XLV. DIETHYLENE GLYCOL DINITRATE

CAS RN 693210, CEE q: 10 t.
Formula — $C_4H_8N_2O_7$ (liquid).
Flammability — There is a high fire hazard when it is exposed to heat or flame. As other nitrates, it is a powerful oxidizing agent.
Explosion — Severe explosion hazard when shocked or exposed to heat. Used in low-freezing dynamites and some permissible explosives.
Hazardous potentials — When heated it emits toxic fumes of NO_x; it can react vigorously with oxidizing or reducing materials.
Toxicity — Moderately toxic via oral route. This compound can cause a dropping blood pressure and possibly various cardiac disturbances.
LD_{50}: 777 mg/kg (administered to rat via oral route).

XLVI. DIALIFOS (DIALIFOR; PHOSPHORO DITHIOIC ACID, *S*-[2-CHLORO-I-(1,3-DIHYDRO-1,3-DIOXO-2H-ISOINDL-2-YL) ETHYL] *O, O*- DIETHYLESTER)

CAS RN 10311849, CEE q: 100 kg.
Formula — $C_{14}H_{17}NO_4PS_2$.
Hazardous potentials — Nonflammable. When heated to decomposition it emits highly toxic fumes of NO_x, PO_x, and SO_x.
Toxicity — Dialifos is highly toxic via oral and dermal routes. It acts as a cholinesterase inhibitor.
TDL_0: 100 mg/kg (administered to hamster via oral route).
LD_{50}: 5 mg/kg (administered to rat via oral route).
LD_{50}: 39 mg/kg (administered to mouse via oral route).
LD_{50}: 94 mg/kg (administered to dog via oral route).
LD_{50}: 35 mg/kg (administered to rabbit via oral route).
LD_{50}: 145 mg/kg (administered to rabbit by skin).

Special precautions and preventive measures — Provide an adequate ventilation. Wear chemical goggles, respirator, and protective clothes (as rubber boots and gloves).

XLVII. DIAZODINITROPHENOL (5,7-DINITRO-1,2,3-BENZOXADIAZOLE)

CAS RN 7008813, CEE q: 10 t.
Formula — $C_6H_2N_4O_5$.
Hazardous potentials — Class A explosive. When heated to decomposition it emits toxic fumes of NO_x.

XLVIII. *O,O*-DIETHYL-*S*-2-ISOPROPYLMERCAPTOMETHYLDITHIOPHOSPHATE

CAS RN 78524, CEE q 100 kg.
Formula — $C_8H_{19}O_2PS_3$.
Hazardous potentials — When heated to decomposition it emits very toxic fumes of PO_x and SO_x.
Toxicity — It is a highly toxic substance by oral and subcutaneous routes of administration.
LD_{50}: 1100 µg/kg (administered to rat via oral route).
LD_{50}: 2 mg/kg (administered to rat via subcutaneous route).
LD_{50}: 7260 µg/kg (administered to chicken via oral route).

XLIX. DIETHYL PEROXYDICARBONATE

CAS RN 14666785, CEE q: 50 t.
Formula — $C_6H_{10}O_6$.
Hazardous potentials — Decompose rapidly (T > 10°C). Do not allow it to crystallize. Sensitive to heat or impact. Powerful explosive.

L. DIISOBUTYRYL PEROXIDE

CAS RN 3437841, CEE q: 50 t.
Formula — $C_8H_{18}O_4$.
Hazardous potentials — Pure material explodes on standing at room temperature.

LI. (DIMETHYLAMINO) CARBONYL CHLORIDE (DMCC)

CAS RN 79447, CEE q: 1 kg.
Formula — C_3H_6ClNO.
Hazardous potentials — When heated to decomposition or on contact with acids or acid fumes it emits highly toxic fumes of chlorides (Cl) and NO_x. It will react with water or steam to produce toxic and corrosive fumes.
Toxicity — Carcinogenic determination: animal positive. It causes skin and papillary tumors in mice due to skin exposure. DMCC is also highly toxic by inhalation. Squamous cell carcinoma in rats due to inhalation.
TCL_0: 1 ppm (administered to rat via inhalation route; equivocal tumorigenic agent; this definition on the base of studies reporting uncertain but seemingly positive results).
TDL_0: 13 mg/kg (administered to mouse by skin over a period of 55 weeks intermittently; neoplastic effects).
TDL_0: 2560 mg/kg (administered to mouse via intraperitoneal route over a period of 64 weeks intermittently; neoplastic effects).

TDL_0: 5200 mg/kg (administered to mouse via subcutaneous route, over a period of 26 weeks intermittently; neoplastic effects).

TD: 8 mg/kg (administered to mouse over a period of 40 weeks, intermittently, in a number of discrete doses; neoplastic effects).

TC: 1 ppm (administered to rat via inhalation route over a period of 6 weeks, intermittently, in a number of discrete doses of 6 h; equivocal tumorigenic agent).

LD_{50}: 1000 mg/kg (administered to rat via oral route).

LCL_0: 1000 mg/m_3 (administered to mouse via inhalation route; duration of exposure 10 min).

Because of the demonstrated carcinogenesis in animals, dimethyl carbamoyl chloride has been considered as an industrial substance suspected of carcinogenic potential for man, with no TLV assigned at this time.

LII. DIMETHYL-2-CHLORO-2-DIETHYLCARBAMOYL-1-METHYLVINYL-PHOSPHATE

CAS RN 13171216, CEE q: 100 kg.

Formula — $C_{10}H_{19}ClNO_5P$.

Hazardous potentials — When heated to decomposition it emits very toxic fumes of Cl, NO_x, and PO_x.

Toxicity — It is a cholinesterase inhibitor; it is highly toxic via oral and dermal routes. LDL_0 (man) 5 mg/kg. Acceptable daily intake: temporary 0.0005 mg/kg body weight.

Special precautions and preventive measures — Corrosive for iron, aluminum, and tin plate. Provide an adequate ventilation. Wear chemical respirators, chemical goggles, rubber gloves, shoes, and other adequate protective clothes.

LIII. 2,4-DIMETHYL-1,3-DITHIOLANE-2-CARBOXALDEHYDE *O*-(METHYLCARBAMOYL)OXIME

CAS RN 26419738, CEE q: 100 kg.

Formula — $C_8H_{14}N_2O_2S_2$.

Hazardous potentials — When heated to decomposition it emits very toxic fumes of NO_x and SO_x.

Toxicity — It is highly toxic via oral and dermal routes.

LD_{50}: 1 mg/kg (administered to rat via oral route).

LD_{50}: 300 mg/kg (administered to rat by skin).

LIV. DIMETHYLNITROSOAMINE (*N*-NITROSODIMETHYLAMINE)

CAS RN 62759, CEE q: 1 kg.

Formula — $C_2H_6N_2O$.

Hazardous potentials — When heated to decomposition it emits toxic fumes of NO_x.

Toxicity — Nitrosodimethylamine is suspected of inducing cancer, based on ether limited epidemiologic evidence. IARC carcinogenic determination: human suspected and OSHA: carcinogen. TLV-TWA: none. Extraordinary care shall be taken both in manufacture and in handling so that worker exposure by all routes is kept below the limit of sensitivity of the analytic method of determining the exposure concentration (NIOSH Manual of Analitical Methods).

LDL_0: 10 mg/kg (administered to woman via oral route over a period of 80 weeks intermittently).

Special precautions and preventive measures — Provide adequate ventilation. Wear masks and working protective clothes, such as rubber gloves, face shields, and chemical goggles.

Disposal and waste treatment — Pour over soda ash. After mixing with the soda ash, drain slowly into a large vessel. Neutralize with 6 *M* HCl and wash into the drain with abundant water.

LV. 2,4-DINITROPHENOL SODIUM SALT HYDRATE

CEE q: 50 t.

Formula — $C_6H_4N_2O_5$.

Hazardous potentials — When heated to decomposition it emits toxic fumes of NO_x.

Toxicity — The harmful effects can be death, damage to the liver, and induced fever. It is a powerful stimulant of metabolism. It is the excessive oxidation effect of this material upon the metabolism and nutrition that damages the liver and kidney cells. It is an irritant to the skin and causes dermatitis.

LDL_0: 20 mg/kg (administered to dog via intravenous route).

LDL_0: 15 mg/kg (administered to pigeon via intravenous route).

LD_{50}: 190 mg/kg (administered to rat).

LD_{50}: 200 mg/kg (administered to mouse).

LDL_0: 1000 mg/kg (administered to dog).

LVI. DIPHACINONE

CAS RN 82666, CEE q: 100 kg.

Formula — $C_{23}H_{16}O_8$.

Toxicity — Diphacinone is highly toxic via oral route; it acts as a cholinesterase inhibitor. It is used as a pesticide. LDL_0: 5 mg/kg (via oral route).

LVII. DI-*n*-PROPYLPEROXYDICARBONATE (CONCENTRATION >80%)

CAS RN 16066389, CEE q: 50 t.

Formula — $C_8H_{14}O_6$.

Hazardous potentials — When heated to decomposition it emits acrid smokes and fumes.

Toxicity — It is a toxic substance via oral and dermal routes. Ingestion or contact with skin may cause moderate erythema, moderate edema.

LD_{50}: 3400 mg/kg (administered to rat via oral route).

LD_{50}: 3500 mg/kg (administered to rabbit by skin).

LVIII. DISULFOTON

CAS RN 298044, CEE q: 100 kg.

Hazardous potentials — When heated to decomposition it emits highly toxic fumes of PO_x and SO_x.

Toxicity — Disulfoton affects human body via inhalation and oral routes. It may be absorbed through skin. It acts as a cholinesterase inhibitor. Disulfoton interfers in the action of mixed function oxidases in the same manner and degree as parathion for which an acceptable TLV has been recommended. The LD_{50} is roughly similar and the indirect action of disulfoton is similar to that of parathion. TLV-TWA: 0.1 mg/m^3 (skin). This value of TLV should serve to limit general room contamination and thereby the cutaneous exposure as well. Acceptable daily intake: 0.002 mg/kg body weight. Maximum residue limits (definition of residue: sum of disulfoton, demeton-S and their sulfoxides and sulfones, expressed as disulfoton):

Commodity	MRL (mg/kg)
Alfalfa (hay)	10
Celery	0.5
Cereal grains (except rice and maize)	0.2
Clover (hay)	10
Coffee beans	0.1
Forage crops (green)	5
Maize	0.5
Peanuts (kernels)	0.1
Pecans	0.1
Pineapples	0.1
Potatoes	0.5
Rice in the husk	0.5
Soya beans (dry)	0.1
Sugar beets (roots)	0.5
Vegetables	0.5

Special precautions and preventive measures — Store in separated, well-ventilated locations. Wear chemical goggles, respirator, other impervious, overall protective clothing (gloves, hats, boots), and accessories to avoid skin contact.

LIX. ETHION (ETHYL METHYLENE PHOSPHORODITHIOATE)

CAS RN 563122, CEE q: 100 kg.

Formula — $C_9H_{22}O_4P_2S_4$.

Hazardous potentials — When heated to decomposition it emits highly toxic fumes of oxides of sulfur and phosphorous.

Toxicity — Ethion is highly toxic via dermal routes. It is very highly toxic via oral and intraperitoneal routes. It may cause central nervous system effects and blood effects (effects on all blood elements). TLV: air 400 μg/m$_3$ (skin). TDL$_0$: 100 μg/kg (via oral route) (effects on all blood elements also on the oxygen carrying or releasing capacity). Ethion is an insecticide, used also as a food additive permitted in the feed and drinking water of animals and/or for the treatment of food-producing animals. Also permitted in food for human consumption.

Special precautions and preventive measures — Wear safety glasses and gas mask. Weak determination of cholinesterase in exposed personnel is recommended.

Disposal and waste treatment — ''(1) Place the disposal in an open furnace. Add to equal parts of sand and crushed limestone, then cover with cumbustible solvent (alcohols, benzene, etc.). Ignite from a safe distance and burn with the utmost care. (2) Place the disposal with sand and crushed limestone into a paper carton. Burn in the furnace with afterburner and alkali scrubber.''[2]

LX. ETHYLENE DIBROMIDE

CAS RN 106934, CEE q: 50 t.

Formula — CH_3BrCH_2Br.

Flammability — Not combustible.

Hazardous potentials — It is a toxic gas and vapors such as hydrogen bromide, bromine, and carbon monoxide may be released when ethylene dibromide decomposes.

Incompatibilities — Ethylene dibromide reacts with chemically active metals such as sodium, potassium, calcium, aluminum, zinc, magnesium, strong alkalis, liquid ammonia, and strong oxidizers.

Toxicity — Ethylene dibromide may affect the body via inhalation, dermal, and oral routes; it may be absorbed through the skin. Ethylene dibromide vapor is a narcotic, a severe mucous membrane irritant, and a hepatic toxin. The liquid is highly irritating to human skin, causing marked erythema and vesiculation. In a bioassay conducted by the National Cancer Institute (NCI) ethylene dibromide was found carcinogenic to rats and mice. It is implicated in worker sterility. OSHA standard air: TWA: 20 ppm averaged over an 8-h work shift; CL: 30 ppm; PK: 50 ppm/5 min during an 8-h work shift. Occupational exposure to ethylene dibromide recommended standard: 1 mg/m^3/15 min (NIOSH). Ethylene dibromide is a suspected human carcinogen.

LDL_0: 90 mg/kg (administered to woman via oral route).

Special precautions and preventive measures — Liquid ethylene dibromide will attack some forms of plastics, rubber, and coatings. Protect containers against physical damage. Store in a cool, dry, well-ventilated location away from any area with fire hazard. Wear safety glasses, gas mask, rubber gloves, and protective work gown. Maintain good ventilation; remove all the workers with diseases of liver and kidney. Provide respirators where it is necessary. Annual medical examinations should be made available to exposed personnel.

Disposal and waste treatment — "Disposal in a combustible solvent. Scatter the spray of the solution into the furnace with after burner and alkali scrubber. Dispose in a secured sanitary landfill."[3]

LXI. ETHYLENE GLYCOL DINITRATE (EGDN)

CAS RN 628966, CEE q: 10 t.

Formula — $C_2H_4N_2O_6$.

Flammability — As other nitrates it presents a moderate fire hazard by spontaneous chemical reaction; it is a powerful oxidizing agent.

Explosion — As other nitrates it may explode when shocked, exposed to heat or flame, or by spontaneous chemical reaction. Explosion point: 114°C.

Hazardous potentials — On decomposition it emits toxic fumes PO_x. As other nitrates it is a powerful oxidizing agent which may cause violent reaction with reducing materials.

Toxicity — Moderately toxic via oral route; when inhaled or absorbed by skin it may cause moderate edema or moderate erythema. It may cause lowered blood pressure, leading to headache, dizziness, and weakness. The physiologic effects of EGDN and those of nitroglycerine are believed to be closely similar. As nitroglycerine, EGDN may cause vasodilation; it affects the heart (dilation of coronary arteries) and the arteries of the dura, and may result in headache and decreases in blood pressure in some individuals; for these effects it also may have therapeutic applications. TLV-TWA: 0.05 ppm (skin). OSHA standard air Cl 1 mg/m^3 (skin) and occupational exposure to nitroglycerine or ethylene glycol dinitrate air CL 0.1 mg/m^3/20 min.

Special precautions and preventive measures — As other nitrates, ethylene glycol dinitrate should be protected carefully.

LXII. ETHYLENE OXIDE

CAS RN 75218, CEE q: 50 t.

Formula — C_2H_4O.

Flammability — At room temperature ethylene oxide is an extremely flammable and reactive gas. Flash point: −17.8°C; flammable limits: 3 to 100%; and ignition temperature: 429°C. The liquid is readily flammable, but not explosive.

Explosion — Severe explosion hazard when exposed to flame. The gas may explode even in the absence of air or oxygen. Explosive limits: 3 to 100% by volume in air.

Hazardous potentials — It reacts violently with many compounds: acids and basis, alcohols, air, *m*-nitroaniline, aluminum chloride, aluminum oxide, ammonia, potassium, bromoethane, magnesium, and copper.

Toxicity — TLV: 1 ppm, weighed over an 8-h shift (adopted in 1984). OSHA: standard air TWA: 50 ppm.

TCL_0: 500 ppm/2 min (administered to woman via inhalation route).

TCL_0: 12500 ppm/10 s (administrated to man via inhalation route; irritant effects).

LC_{50}: 4000 ppm/1 h (administrated to man via inhalation route).

Aqueous solutions of ethylene oxide are extremely irritating to the skin with effects often appearing after a latent period of 1 to 5 h. Ethylene oxide is absorbed by rubber, plastic, and leather articles during sterilization. Rubber gloves and shoes sterilized with ethylene oxide have caused severe skin irritation. Ethylene oxide is suspected of carcinogenic potential for man.

Special precautions and preventive measures — Protect containers against physical damage and check leakage intermittently. Store in distant outdoor tank (or container) protected from direct sunlight, lined with insulating material, and equipped with adequate refrigerator system and water sprinkler system (T < 30°). Indoor storage should be restricted to as small quantitites as possible and to the combustible liquid storage cabin that is fireproof in conformity with regulations. Provide adequate ventilation and store away from other combustibles, other oxides, chlorides, acids, organic basis, caustic alkalis, and metallic sodium. Spark-resistant design is preferable. Wear chemical goggles, neoprene gloves, protective clothing, and self-contained breathing apparatus. Particular attention should be made that one can come in contact with the compound wearing contaminated clothes; the compound can saturate rubber clothing.

Disposal and waste treatment — (1) Place on ground in an open area. Evaporate or burn by igniting from a safe distance, or (2) dissolve in benzene, petroleum ether, or higher alcohol such as butanol. Dispose by burning the solution.

LXIII. ETHYLENIMINA

CAS RN 151564, CEE q: 50 t.

Formula — C_2H_5N.

Flammability — Flammable even at room temperature. Dangerous when exposed to heat, flame, or oxidizers. The heavier than air vapor sometimes travels considerable distance to sources of ignition and flashback. The concentrate violently polymerizes in contact with acids. At elevated temperature such as in fire conditions the substance may polymerize with evolution of heat and the possibility of a violent rupture of the container. flash point: − 13°C; ignition temperature 322°C; and flammable limits: 3.6 to 46% by volume in air.

Hazardous potentials — Reacts violently with acetic acid, acrolein, CS_2, Cl_2, HCL, Hf, HNO_3, H_2SO_3, acetic anhydride, and acrylic acid. Heat and/or the presence of catalytically active metals may cause a violent exothermic reaction.

Toxicity — Perceptible odor level: 2 ppm; TLV-TWA: 0.5 ppm (skin); OSHA: carcinogen; IARC carcinogenic determination: animal positive. OSHA standard air: TWA: 1 mg/m³. It is a highly irritating substance for skin and eyes, via oral, dermal, percutaneous, and inhalation routes. It is an allergic sensitizer of skin. It has been known to cause human eye injury (cornea). Inhalation of ethylenimine results in dalayed lung injury with congestion, edema and hemorrage, and kidney damage. It may cause a decrease in white blood cell count.

Special precautions and preventive measures — Provide adequate ventilation. Protect containers against physical damage. Outside or detached storage is preferred. For indoor storage use standard combustible liquid storage room. Avoid fire and direct sunlight. Wear butyl rubber gloves, self-contained breathing apparatus, and plastic working clothing (apron, boots); rubber may saturate with Ethylenimine.

Disposal and waste treatment — (1) Dissolve in such combustible solvent. Spray the solution into the furnace with after burner and scrubber or (2) pour into a mixture of sand soda ash. Burn in the furnace with paper to serve as fuel.

LXIV. ETHYL ETHER (DIETHYL ETHER)

CAS RN 60297, CEE q.
Formula — $(C_2H_5)_2O$.
Flammability — Very dangerous when exposed to heat or flame. It may travel considerable distance to the source of ignition to cause a flashback. It may be ignited by static electricity, spark, or impact. Flash point: $-45°C$; ignition temperature: $160°C$; and flammable limits: (percent by volume) lower: 1.9 and upper: 36.

Hazardous potentials — Toxic gases and vapors (such as carbon monoxide) may be released in a fire involving ethyl ether. Ethers, which have been in contact with air or exposed to light for a long time, may contain unstable peroxides which can explode spontaneously, or when heated, or when the caps or stoppers of their containers are removed. Ethyl ether is a nonconductor and it may accumulate static electric charges which may result in an ignition source. It may explode or cause fires in contact with oxidizing agents.

Toxicity — Ethyl ether is a depressant of the central nervous system and it is capable of producing intoxication, unconsciousness; it has predominantly narcotic properties. It is an irritant to the eyes, nose, and throat. Concentration anesthetic to human subjects range from 3.6 to 6.5 vol % in air, respiratory arrest from 7 to 10 (volume percent in air); concentrations greater than 10 (volume percent in air) are fatal. TLV-TWA: air 400 ppm; TLV-STEL: 500 ppm; and OSHA standard: air TWA 400 ppm.

LDL_0: 420 mg/kg (administered via oral route).

LDL_0: 260 mg/kg (administered via oral route).

TCL_0: 200 mg/kg (administered via inhalation route).

Acquatic toxicity rating: TLm_{96} over 1000 ppm.

Special precautions and preventive measures — Protect containers against physical damage. Detached outdoor storage is preferred. Indoor storage should be in a standard flammable liquid storage room. Avoid direct sunlight. Protect against static electricity and lightning. The floors of the storage room should be conductive. Wear chemical goggles, chemical cartridge respirator, and rubber gloves. The shoes should be conductive and nonsparking. The protective clothing should be made from a special rubber compound, resistant to swelling by ether. Store away from other combustible materials, explosives, toxical, radioactive, oxidizing materials, or organic peroxides.

Disposal and waste treatment — "(1) Place on ground in an open area. Evaporate by igniting from a safe distance or (2) dissolve in benzene, petroleum ether, or higher alcohol, such as butanol. Dispose by burning the solution."[2]

LXV. ETHYL-γ-FLUOROBUTYRATE

CAS RN 63904961, CEE q: 1 kg.
Formula — $C_6H_{11}FO_2$.
Hazardous potentials — When heated to decomposition it emits toxic fumes of F.
Toxicity — Highly toxic via inhalation route.

LCL_0: 500 mg/m³ (administrated to rat via inhalation route; dermation of exposure: 10 min).

LCL_0: 500 mg/m³ (administered to mouse via inhalatiron route; duration of exposure: 10 min).

LCL_0: 500 mg/m³ (administered to guinea pig via inhalation route; duration of exposure: 10 min).

LXVI. ETHYL GUTHION

CAS RN 2642719, CEE q: 100 kg.
Formula — $C_{12}H_{16}N_3O_3PS_2$.
Hazardous potentials — When heated to decomposition it emits highly toxic fumes of SO_x, PO_x, and NO_x.
Toxicity — Ethyl guthion is very toxic via oral and intraperitoneal route and highly toxic via dermal route. It is a cholinesterase inhibitor.
LD_{50}: 9 mg/kg (administered to rat via oral route).
LD_{50}: 250 mg/kg (administered to rat by skin).
LD_{50}: 7500 mg/kg (administered to rat via intraperitoneal route).
LD_{50}: 34 mg/kg (administered to chicken via oral route).

LXVII. ETHYL NITRATE

CAS RN 625581, CEE q: 50 t.
Formula — $C_2H_5NO_3$.
Flammability — Dangerous when exposed to heat or flame. No spontaneous heating.
Explosion — Moderate explosion hazard when exposed to heat.
Hazardous potentials — On decomposition it emits toxic fumes. As other nitrates it is a powerful oxidizing agent which may cause violent reaction with reducing materials. Incompatibilities: Lewis acids.
Toxicity — Probably moderately toxic via oral and inhalation routes. See Section CXL. Aquatic toxicity rating TLm_{96} over 1000 ppm.

LXVIII. ETHYL PYRAZINIL PHOSPHOROTHIOATE

CAS RN 297972, CEE q: 100 kg.
Formula — $C_8H_{13}N_2O_3PS$.
Hazardous potentials — Thionazin is a dangerous substance; when heated to decomposition it emits highly toxic fumes (NO_x, PO_x, and SO_x).
Toxicity — It is a cholinesterase inhibitor (see also Section CXII). It is highly toxic via oral and dermal routes.
LD_{50}: 3500 μg/kg (administered to rat via oral route).
LD_{50}: 8 mg/kg (administered to rat by skin).
LD_{50}: 10 mg/kg (administered to guinea pig by skin).
LD_{50}: 1680 μg/kg (administered to duck via oral route).
LD_{50}: 7 mg/kg (administered to duck by skin).
Special precautions and preventive measures — Wear chemical respirator, chemical goggles, rubber gloves and shoes, and other overall protective clothing. Provide an adequate ventilation; protect containers against physical damage.

LXIX. FENSULFOTHION: *O,O*-DIETHYL *O*-(*P*-(METHYLSULFINYL) PHENYL)PHOSPHOROTHIOATE

ENT 24, 945, CAS RN 115902, CEE q: 100 kg.
Formula — $C_{11}H_{17}O_4PS_2$.

Hazardous potentials — When heated to decomposition it emits very toxic fumes of SO_4 and PO_x.

Toxicity — Highly toxic via oral, inhalation, dermal, intraperitoneal routes. It is a cholinesterase inhibitor. TLV-TWA: air 0.1 mg/m^3 (skin). Acceptable daily intake: 0.3 μg/kg body weight.

Commodity	MRL (mg/kg)
Bananas	0.02
Cattle, carcass meat (in the carcass fat)	0.02
Cattle, edible offal	0.02
Goats, carcass meat (in the carcass fat)	0.02
Goats, edible offal	0.02
Maize (grain, including kernels of field corn and popcorn	0.1
Onions	0.1
Peanuts (on a shell-free basis)	0.05
Pineapples	0.05
Potatoes	0.1
Sheep, carcass meat (in the carcass fat)	0.02
Sheep, edible offal	0.02
Sugar beets	0.1
Swede (rutabaga roots)	0.1
Tomatoes	0.1

Special precautions and preventive measures — Provide adequate ventilation. Wear chemical goggles, chemical respirator, other overall protective clothes, rubber shoes, and gloves.

LXX. FLUOROACETIC ACID AMIDE

CAS RN 640197, CEE q: 1 kg.
Formula — C_2H_4FNO.
Hazardous potentials — When heated to decomposition it emits very toxic fumes of F and NO_x.
Toxicity — It is highly toxic via oral, dermal, intraperitoneal, or subcutaneous routes.
TDL_0: 90 mg/kg (administered to rat via oral route, during a period of 50 d).
LD_{50}: 5750 μg/kg (administered to rat via oral route).
LD_{50}: 80 mg/kg (administered to rat by skin).
LD_{50}: 31 mg/kg (administered to mouse via oral route).
LD_{50}: 85 mg/kg (administered to mouse via intraperitoneal route).
LD_{50}: 34 mg/kg (administered to mouse via subcutaneous route).
LD_{50}: 250 μg/kg (administered to rabbit via intravenous route).

LXXI. FLUOROETHANOIC ACID

CAS RN 144490, CEE q: 1 kg.
Formula — $C_2H_3FO_2$.
Hazardous potentials — When heated to decomposition it emits toxic fumes of F.
Toxicity — It may cause central nervous system effects: convulsions and ventricular fibrillation. It is highly toxic via subcutaneous, intraperitoneal, or intravenous route.

TDL$_0$: 2 mg/kg (via oral routes: central nervous system effects).

LXXII. 4-FLUORO-2 HYDROXY BUTYRIC ACID SODIUM SALT: SODIUM γ-FLUORO-β-HYDROXY BUTYRATE

CEE q: 1 kg.
Formula — C$_4$H$_6$FO$_3$Na.
Hazardous potentials — When heated to decomposition it emits toxic fumes of F and Na$_2$O.
Toxicity — It is highly toxic via oral and subcutaneous routes.
LDL$_0$: 1 mg/kg (administered to rat via oral route).
LDL$_0$: 500 μg/kg (administered to mouse via subcutaneous route).

LXXIII. γ-FLUORO-β-HYDROXY BUTYRIC ACID THIOMETHYL ESTER

CEE q: 1 kg.
Formula — C$_5$H$_9$FO$_2$S.
Hazardous potentials — When heated to decomposition it emits very toxic fumes of SO$_x$ and F.
Toxicity — It is a highly toxic substance via inhalation route. It may cause moderate erythema and moderate edema.
LCL$_0$: 30 mg/m^3/10 min (administered to mouse via inhalation route.
LCL$_0$: 63 mg/m^3/10 min (administered to dog via inhalation route).
LC$_{50}$: 188 mg/m^3/10 min (administered to monkey via inhalation route).
LCL$_0$: 63 mg/m^3/10 min (administered to rabbit via inhalation route).
LCL$_0$: 63 mg/m^3/10 min (administered to cat via inhalation route).

LXXIV. FORMALDEHYDE

CAS RN 50000, CEE q: 50 t.
Formula — CH$_2$O.
Flammability — Flammable limits: lower: 7.0% and upper 73%; flash point: 66°C; and ignition temperature: 300°C. It is flammable in air. It will burn above flash point if exposed to flame and sparks.
Explosion — When aqueous formaldehyde solutions are heated above their flash points, a potential for explosion hazard exists. Reacts with NO$_x$ at about 180°C. The reaction becomes explosive. It is highly reactive and readly polymerizes with various organic materials.
Toxicity — Formaldehyde is highly irritant to skin, eyes, and mucous membrane. It is highly toxic; it is a suspected human carcinogen. TLV-TWA: 1 ppm: OSHA standard air TWA: 3 ppm: CL: 5 ppm: PK 10 ppm/30 min/8-h work shift. Occupational exposure to formaldehyde recommended standard: air: CL: 1.2 mg/m^3/30 min.
LDL$_0$: 36 mg/kg (administered to woman via oral route).
TCL$_0$: 17 mg/m^3/30 min (administered via inhalation route).
Special precautions and preventive measures — Protect containers against physical damage. Separate from oxidizing and alkaline materials. Indoor storage should be in areas having floors pitched toward a trapped drain or in curbed retention areas. Store where temperature range is 16 to 35°C. Formaldhyde should not be stored in confined spaces or near open flames. Indoor storage areas should be equipped with automatic sprinklers. Storage tanks should be adequately grounded to discharge static electricity and to reduce other electrical hazard. Wear chemical goggles, chemical cartridge respirator or airline mask, and rubber protective clothing. Provide adequate ventilation and frequent examinations of exposed personnel for early signs of skin irritation.

Disposal and waste treatment — Dissolve in a combustible solvent, then spray the solution into a furnace with afterburner.

LXXV. HEXACHLORODIBENZO-*P*-DIOXIN (HCDD)

CAS RN 34465468.
Formula — $C_{12}H_2CL_6O_2$.
Toxicity — HCDD is highly toxic via oral route. It is an equivocal tumorigenic agent; this definition on the basis of studies reporting uncertain but seemingly positive results.
TDL_0: 1 μg/kg (administered to rat via oral route over a period of 6 to 15 d).
TDL_0: 10 μg/kg (administered to rat via oral route over a period of 6 to 15 d).
TCL_0: 1 mg/kg (administered to rat via oral route over a period of 6 to 15 d).
LDL_0: 100 mg/kg (administered to rat via oral route).
LD_{50}: 100 mg/kg (administered to rat via oral route).
LD_{50}: 1250 μg/kg (administered to mouse via oral route).
LD_{50}: 70 μg/kg (administered to guinea pig via oral route).

LXXVI. 1,2,3,6,7,8-HEXACHLORODIBENZO-*P*-DIOXIN MIXED WITH 1,2,3,7,8,9 HEXACHLORODIBENZO-*P*-DIOXIN (67%)

CAS RN 34465468, CEE q: 100.
Hazardous potentials — When heated to decomposition it emits very toxic fumes of Cl and dioxin.
Toxicity — Highly toxic by oral route. It is an equivocal tumorigenic agent; this definition on the basis of studies reporting uncertain but seemingly positive results.

LXXVII. HEXAMETHYLPHOSPHOROTRIAMIDE (HEXAMETHYL PHOSPHORAMIDE)

CAS RN 680319, CEE q: 1 kg
Formula — $C_6H_{18}N_3OP$.
Hazardous potentials — When heated to decomposition it emits very toxic fumes of PO_x and NO_x.
Toxicity — Hexamethyl phosphoramide is a suspected human carcinogen. Administration via oral, dermal, or intravenous route may cause moderate erythema and moderate edema. Carcinogenic determination: animal positive (IARC).
LD_{50}: 2525 mg/kg (administered to rat via oral route).
LDL_0: 3500 mg/kg (administered to rat by skin).
LD_{50}: 800 mg/kg (administered to mouse via intravenous route).
LD_{50}: 2600 mg/kg (administered to rabbit by skin).
LD_{50}: 1600 mg/kg (administered to guinea pig via oral route).
LD_{50}: 1175 mg/kg (administered to guinea pig by skin).
LD_{50}: 835 mg/kg (administered to chicken via oral route).
TCL_0: 440 ppb (administered to rat via inhalation route over a period of 35 weeks, intermittently, in a number of discrete doses). No TLV is recommended at this time pending the completion of experiments defining a no-observed effect level. It appears that a TLV should be in the low ppb range, at best.

LXXVIII. 2,4,6,2′,4′,6′,- HEXANITRODIPHENYLAMINE

CAS RN 131737, CEE q: 50 t.
Formula — $C_{12}H_5N_7O_{12}$.

Flammability — It is flammable, as other nitrates. it is a powerful oxidizing material, a powerful and violent explosive used as a booster explosive, in which use it is superior to TNT. It is not as good for this purpose as tetryl, but is extremely stable and much safer to handle.

Hazardous potentials —On decomposition it emits toxic fumes of NO_x.

Toxicity — Highly toxic via oral route.

TDL_0: 14 mg/kg (administered to rat via oral route over a period of 76 weeks).

LCL_0: 250 mg/kg.

LXXIX. HYDROCYANIC ACID (HYDROGEN CYANIDE)

CAS RN 74908, CEE q: 20 t.

Formula — HCN.

Flammability — Flammable and dangerous fire hazard; flash point $-18°C$; ignition temperature: 538°C; and flammable limits: 6 to 41% by volume in air.

Explosion — Severe explosion hazard when hydrogen cyanide is exposed to heat or flame or when it reacts with oxidizers. It may cause explosion by contact with alkalis.

Hazardous potentials — The gas forms explosive mixtures with air; it will react with water, steam, acid, or acid fumes to produce highly toxic fumes of cyanides. Decomposition products are less toxic than hydrogen cyanide itself. On long standing, liquid hydrogen cyanide polymerizes to form a dark brown explosive solid.

Toxicity — Hydrogen cyanide can affect the body if it is inhaled, if it comes in contact with the eyes or skin, or if swallowed. It may be absorbed through skin. Hydrogen cyanide (HCN) vapor, a source of cyanide ion, is an asphyxiant due to an inhibitory action on metabolic enzyme system and can be rapidly fatal. Cyanide exerts this effect because it inactivates certain enzymes by forming very stable complexes with the metal in them; these complexes render the oxygen unavailable to the tissues. At 270 ppm HCN is immediately fatal to humans, 181 ppm is fatal after 10 min, 135 ppm after 30 min; and 110 ppm may be fatal in 1 h. The ingestion by humans of 50 to 100 mg of HCN may also be fatal. TLV: 10 ppm (ceiling value); OSHA standard air: TWA 10 ppm (skin); Occupational exposure to hydrogen cyanide recommended standard CL: 5 mg/m².

LDL_0: 570 μg/kg (via oral route).

LCL_0: 120 mg/m³/1 h (via inhalation route).

LCL_0: 200 mg/m³/10 min (via inhalation route).

LDL_0: 1 mg/kg (via subcutaneous route).

LD_{50}: 1 mg/kg (via intravenous route).

Special precautions and preventive measures — Protect containers against physical damage. Remove every source of ignition. Outdoor or detached storage is preferred. Separate from other storages especially from combustible materials. Wear long rubber gloves, self-contained breathing apparatus, aprons, and special impervious clothing.

Disposal and waste treatment — Seal the cylinder and return to suppliers.

LXXX. HYDROFLUORIC ACID (HYDROGEN FLUORIDE)

CAS RN 7664393, CEE q.

Formula — HF.

Flammability — It is not combustible.

Hazardous potentials — Liquid or gaseous hydrogen fluoride will attack metals, concrete, glass, ceramics, any silica-containing materials, plastics, and rubber coatings. It does not corrode lead, steel, and platinum, (and polyethylene and polystyrene if diluted). May generate metal fluorides and hydrogen in contact with some metals. When heated it emits highly corrosive fumes of fluorides. Will react with water or steam to produce toxic and corrosive fumes.

Toxicity — It is highly irritating to skin, eyes and mucous membrane and via oral route; hydrogen fluoride as a gas is a severe respiratory irritant and in solution it causes severe and painful burns of the skin. TLV = 3 ppm (ceiling limit); OSHA standard air TWA = 3 ppm. Occupational exposure to HF recommended standard: TWA = 2.5 mg (F)/m³; and CL = 5 mg (F)/m³/15 min.

TCL_0: 110 ppm/1 min (administered to man, irritant effects).

LCL_0: 50 ppm/30 min.

TCL_0: 32 ppm (irritant effects).

Special precautions and preventive measures — Protect against physical damage. Store in well-ventilated area, separated from other storage. Wear neoprene gloves, full face shield, chemical cartridge respirator, and plastic coveralls (aprons, boots etc.).

Disposal and waste treatment — ''Add slowly to a large amount of solution of soda ash and slaked lime by stining. Discharge the solution with large amount of water into a sink lined with protective matting and filled with chipped marble.''[2]

LXXXI. HYDROGEN

CAS RN 1333740, CEE q: 50 t.

Formula — H_2.

Flammability — Hydrogen is a flammable gas. Hydrogen is highly dangerous when exposed to heat, flame, or oxidizers; it is flammable or explosive when mixed with air, O_2, or chlorine. Ignition point: 585°C.

Explosion — Hydrogen presents a severe explosion hazard when exposed to heat or flame. It reacts violently with (air + Pt), Br_2, Cl_2, I_2, ClF_3, NF_3, OF_2, (Pd + isopropyl alcohol), 3-methyl-2-penten-4-yn-1-ol, PbF_3, oxidants and PbF_3, oxidants.

Hazardous potentials — See Explosion.

Toxicity — Hydrogen is harmless in itself. In high concentration it can act as a simple asphyxiant; it may lower contents of oxygen in air upon leakage. Liquid hydrogen may cause serious burnings.

Special precautions and preventive measures — Prohibit fire. The cylinders should be protected against direct sunlight and stored in a well-ventilated area. Protect containers against physical damage. Do no load together with explosives, poisons, radioactive materials, and organic peroxides. Wear safety glasses. Store away from cylinders of oxygen, chlorine, and other halogens. Do not use compressed air to fill, to release, or to handle hydrogen gas. Before introducing hydrogen, containers should be filled with an inert gas to avoid the formation of explosive mixtures. Hydrogen may be stored in steel containers at 150 atm and at temperature lower than boiling temperature.

LXXXII. HYDROGEN CHLORIDE (HYDROCHLORIC ACID)

CAS RN: 7647010, CEE q: 250 t.

Formula — HCl.

Flammability — Nonflammable gas.

Explosion — No explosion risk.

Incompatibilities — Hazardous reaction with ethylene. Contact of hydrogen chloride with most metals corrodes them severly and forms flammable hydrogen gas. Contact of hydrogen chloride gas or liquid with any alkali or active metal may develop enough heat to cause fire in adjacent combustible material.

Toxicity — Hydrogen chloride gas irritates the eyes, mucous membranes, and skin. A concentration of 35 ppm causes irritation of the throat after short exposure. Concentrations of 50 to 100 ppm are tolerable for 1 hr; 1000 to 2000 ppm are dangerous, even for brief exposure; 10 ppm is the maximal concentration allowable for prolonged exposure. OSHA standard air: CL 5 ppm and TLV: 5 ppm (ceiling value).

Preventive measures — Store away from oxidizing materials. Provide an adequate ventilation; wear chemical goggles, chemical cartridge respirator, rubber gloves, and protective clothing. Periodical medical examination for exposed personnel should be made available. Hydrogen chloride will attack most metals and some forms of plastics, rubber, and coatings.

Disposal and waste treatment — Add slowly to a large amount of solution of soda ash and slaked lime by stirring. Discharge the solution with large amount of water into a sink lined with protective matting and filled with chipped marble.

LXXXIII. HYDROGEN SELENIDE

CAS RN 7783075, CEE q: 10 kg.

Formula — H_2Se.

Flammability — Flammable; it will react vigorously with powerful oxidizing agents, such as H_2O_2 and HNO_3.

Hazardous potentials — It forms explosive mixtures with air. Contact with acids, water, and halogenated hydrocarbons may cause fires and explosion (they can liberate hydrogen). Toxic gases and vapors (such as selenium dioxide fumes may be released in a fire involving hydrogen selenide.

Toxicity — Very highly irritant substance to skin, eyes and mucous membrane, and via inhalation route. It is an allergen. It can cause damage to the lungs and liver as well as conjunctivitis. Very little data are available on possible chronic effects. TLV: air 0.05 ppm and OSHA standard air: TWA: 0.05 ppm averaged over an 8-h work shift.

TCL_0: 0.2 ppm (via inhalation route; central nervous system effects).

Special precautions and preventive measures — Usually stored at room temperature, in dry locations. Precautions must be taken to avoid contact with water or prolonged exposure to air and other materials with which it may react. Used drums should be stored in the same area to avoid contamination and to minimize any hazard due to residues. Wear chemical goggles and mechanical filter respirator and other personal protective equipment (such as rubber or plastic gloves). Provide, where it is necessary, local exhaust ventilation system and physical examinations of exposed personnel every 6 months including determinations of selenium in urine and studies of liver function.

Disposal and waste treatment — Burn in a suitable combustion chamber equipped with an appropriate effluent gas cleaning device. Seal cylinder and return to the suppliers.

LXXXIV. HYDROGEN SULFIDE

CAS RN 77830604, CEE q: 50 t.

Formula — H_2S.

Flammability — It is a very dangerous gas when exposed to heat, flame or oxidizers. Autoignition temperature: 260°C and flammable limits in air (per cent by volume): lower 4 to 3% and upper 46. May travel a considerable distance to a source of ignition and flash back.

Explosion — Moderate explosion risk when exposed to heat or flame. It is dangerously reactive (fires and explosions) with fuming or strong nitric acid and powerful oxidizing materials.

Hazardous potentials — When heated to decomposition it emits highly toxic fumes of oxides of sulfur. Hydrogen sulfide attacks many metals, which results in the formation of sulfides.

Toxicity — In high concentrations (500 to 1000 ppm) hydrogen sulfide acts primarily as a systemic poison, causing unconsciousness and death through respiratory paralysis. In lower concentrations (50 to 500 ppm) hydrogen sulfide acts primarily as a respiratory irritant. It is reported that pulmonary edema and bronchial pneumonia may follow prolonged exposure at concentrations exceeding 250 ppm. At low concentrations the effects on the eye predominate, with conjuntivitis the most common effect, while keratitis frequently occurs (20 ppm). TLV-TWA: 10 ppm; TLV-STEL: 15 ppm; and OSHA standard air: CL: 20 ppm and PK: 50 ppm/10 min. Occupational exposure to hydrogen sulfide recommended standard air: 15 mg/m³/10 min. NIOSH has recommended a 10-min ceiling value of 10 ppm and the work areas, in which the concentration of hydrogen sulfide exceeds 70 mg/mg₃, to be evacuated.

Special precautions and preventive measures — The majority of occupational exposure to H_2S however, has resulted from its occurrence in petroleum, natural gas, soil, sewer gas, and as a by-product of chemical reactions. Store in stainless steel (nickel-chrome 316 18-8) containers. Protect containers against physical damage. Outdoor or detached storage is preferred. Indoor storage should be in a cool, well-ventilated, noncombustible location, away from all possible sources of ignition. Store away from nitric acid, strong oxidizing materials, corrosive liquids or gases, cylinders, or other containers under high pressure and possible sources of ignition. Protect against static electricity, direct sunlight, and excessive heat. Wear chemical goggles, chemical cartridge respirator, and rubber gloves.

LXXXV. HYDROXYACETONITRILE

CAS RN 107164, CEE q: 100 kg.
Formula — C_2H_3NO.
Hazardous potentials — When heated to decomposition it emits toxic fumes of NO_x and CN.
Toxicity — It is highly toxic via oral, inhalation, intraperitoneal, subcutaneous, and dermal routes. Occupational exposure to nitriles recommended standard: air CL 5 mg/m³/15 min.

LXXXVI. 5-HYDROXY-1,4-NAPHTHOQUINONE (5-HYDROXY-1, 4-NAPHTHALENEDIONE)

CAS RN 481390, CEE q: 100 kg.
Formula — $C_{10}H_6O_3$.
Hazardous potentials — When heated to decomposition it emits acrid smoke and fumes.
Toxicity — Highly toxic via oral route. An experimented neoplastic agent (NEO).
TDL_0: 394 mg/kg (administered to mouse by skin over a period of 53 weeks intermittently, in a number of discrete doses; neoplastic effects).
LD_{50}: 2500 μg/kg (administered to mouse via oral route).

LXXXVII. ISOBENZAN 1,3,4,5,6,7,8,8-OCTACHLORO-1,3,3a,4,7,7a-HEXAHYDRO-4,7-METHANOISOBENZOFURAN

CAS RN 297789, CEE q: 100 kg.
Formula — $C_9H_4Cl_8O$.
Hazardous potentials — When heated to decomposition it emits toxic fumes of Cl.
Toxicity — Highly toxic via oral, dermal, intraperitoneal, or intravenous routes. It is an experimented equivocal tumorigenic agent.
LDL_0: 6200 μg/kg (administered to rat via oral route).
LD_{50}: 5 mg/kg (administered to rat by skin).

LD_{50}: 3560 μ/kg (administered to rat via intraperitoneal route).
LD_{50}: 8400 μg/kg (administered to mouse via oral route).
LD_{50}: 8170 μg/kg (administered to mouse via intraperitoneal route).
LD_{50}: 1800 μg/kg (administered to rat via intravenous route).
LD_{50}: 1 mg/kg (administered to dog via oral route).
LD_{50}: 5 mg/kg (administered to cat via oral route).
LD_{50}: 4 mg/kg (administered to rabbit via oral route).
LD_{50}: 12 mg/kg (administered to rabbit by skin).
LD_{50}: 2000 μg/kg (administered to guinea pig via oral route).
LD_{50}: 2 mg/kg (administered to guinea pig by skin).
LD_{50}: 7500 μg/kg (administered to hamster via oral route).

LXXXVIII. ISODRIN

CAS RN 465736, CEE q: 100 kg.
Formula — $C_{12}H_8Cl_6$.
Hazardous potentials — As other chlorides, when heated to decomposition or on contact with acids or acids fumes, highly toxic chloride fumes can evolve.
Toxicity — It is highly toxic via oral, dermal route. It causes liver injury, acne, and skin rashers.
LD_{50}: 7 mg/kg (administered to rat via oral route).
LD_{50}: 23 mg/kg (administered to rat by skin).
LD_{50}: 8800 μg/kg (administered to mouse via oral route).
LD_{50}: 6400 mg/kg (administered to mouse vial intraperitoneal route).

LXXXIX. ISOPROPYL-γ-FLUOROBUTYRATE (4-FLUROBUTYRIC ACID, ISOPROPYL ESTER)

CAS RN 63904972, CEE q: 1kg.
Formula — $C_7H_{13}FO_2$.
Hazardous potentials — When heated to decomposition it emits toxic fumes of F.
Toxicity — It is a highly toxic substance via inhalation route.
LCL_0: 10 mg/m³/10 min (administered to mouse via inhalation route).
LCL_0: 100 mg/m³/10 min (administered to guinea pig via inhalation route).

XC. LEAD AZIDE (II)

CAS RN 13424469, CEE q: 50 t.
Formula — N_6Pb.
Explosion — Class A explosive. Severe explosion hazard when shocked or exposed to heat or flame. Explodes at 250°C. Will explode spontaneous during crystallization.
Hazardous potentials — Highly dangerous, shock and heat will explode it; when heated to decomposition it emits highly toxic fumes of lead NO_x. It is incompatible with calcium stearate, copper, or zinc; brass; and CS_2.
Toxicity — Occupational exposure to inorganic lead recommended standard: air: TWA 0.10 mg (Pb)/m³ (lead damage: see Section XCI).

XCI. LEAD TRINITRORESORCINATE

CAS RN 15245440, CEE q: 50 t.
Formula — $C_6HN_3O_8Pb$.

Flammability — High fire hazard. As other nitrates it is a powerful oxidizing agent.

Explosion — Severe explosion hazard when heated. it explodes at 311°C. Class A explosive.

Hazardous potentials — It emits very toxic fumes of NO_x and Pb when heated to decomposition. It is shock sensitive and has detonated spontaneously when dry.

Toxicity — Occupational exposure to inorganic lead recommended standard air: TWA: 0.10 mg (Pb)/m³. Lead is a cumulative poison; it produces a brittleness of the red blood cells so that they hemolyze with but slight trauma; the hemoglobin is not affected. Lead produces a damaging effect on the organs or tissues with which it comes in contact. Lead may enter the body by inhalation of dusts, fumes, mists and vapors, by ingestion of lead compounds trapped in the upper respiratory tract or introduced into the mouth, and through the skin.

XCII. MERCURY FULMINATE

CAS RN 628864, CEE q: 10t.

Formula — $C_2HgN_2O_2$.

Flammability — Dangerous substance. Keep away from heat and open flame.

Explosion — Severe explosion hazard when shocked or exposed to heat, flame, or friction. It is a class A explosive. It should be kept moist until used (it explodes when dried).

Hazardous potentials — When heated to decomposition it emits highly toxic fumes of NO_x and Hg.

Toxicity — Occupational exposure to inorganic mercury recommended standard air TWA: 0.05 mg (Hg)/m³. Mercury is a general protoplasmic poison. In industrial poisoning the principal effect is upon the central nervous system and upon the mouth and gums. Fulminate of mercury rarely produces symptoms of systemic poisoning, but frequently causes dermatitis which take the form of small, discrete ulcers on the exposed parts, and which are usually accompanied by conjunctivitis and inflammation of the mucous membrane of the nose and throat. Spilled and heated elemental Hg is particularly hazardous.

Special precautions and preventive measures — Keep moist until used. Wear rubber gloves, self-contained breathing apparatus, and coveralls. Do not contaminate air and water with this chemical. If compressed beyond 25,000 psi fulminate becomes what is known as "dead-pressed", i.e., not capable of being exploded by flame. As other fulminates it is subject to deterioration when stored in hot climates. All precautions required for protection of magazines apply to storage of this material. As other fulminate it should not be handled when frozen. Wet fulminate of mercury or wet floor coverings containing small quantities of fulminates may be burned on windrows of flammable material. Care is required to prevent fulminate dust from being carried off in the exhaust system; deposits thus made have caused explsions. The floors on which fulminates are used, should be covered with 1/16 in cloth inserted rubber packing or its equal. All cracks and crevices should be covered. The walls of these rooms should be covered with glazed, waterproof material.

Disposal and waste treatment — "Dissolve in water after converting soluble nitrate if the compound is not water soluble. Adjust the pH and precipitate mercury as mercury sulfide. Wash and dry the precipitate and return to the suppliers."[2]

XCIII. 2-METHYLAZIRIDINE

CAS RN 75558, CEE q: 50 t.

Formula — C_3H_7N.

Flammability — Moderate fire hazard when exposed to heat or flame. Flammable liquid. Flash point: −4°C (25°F) (closed cup).

Hazardous potentials — When heated to decomposition it emits toxic fumes of CO and

NO_x. Contact with acids will cause violent polymerization that may burst containers. Contact with strong oxidizers may cause fires and explosions.

Toxicity — Propilene imine is a suspected human carcinogen, affecting a wide variety of organs. Carcinogenic determination: animal positive (IARC). It is an eye irritant substance. Highly toxic via oral, inhalation, or dermal route implicated as a brain carcinogen. TLV-TWA: 2 ppm (skin) and OSHA standard: air: TWA 2 ppm (skin) averaged over a 8-h work shift.

Special precautions and preventive measures — Liquid propylenimine will attack some forms of plastics, rubber, and coatings. Wear impervious clothing, gloves, face shields (8 in. minimum), and other appropriate protective clothing necessary to prevent skin contact with liquid propylene imine.

XCIV. METHYL-ETHYL-KETONE-PEROXIDE

CAS RN 1338234, CEE q: 50 t.

Formula — $C_8H_{16}O_4$.

Hazardous potentials — When heated to decomposition it emits acid smokes and fumes. Severe explosion or fire risk when heated, or shocked, or on contact with reducing agents, organic materials, strong acids, and bases. Methyl-ethyl-ketone-peroxide is sold commercially as a liquid mixture of approximately 60% MEKP and 40% diluent to reduce its sensitivity to shock; these diluents provide a stable mixture which under ordinary handling cannot be detonated at room temperature. Separation of shock-sensitive MEKP from these diluents can create a hazardous situation.

Toxicity — TLV: 0.2 ppm and TDL_0: 480 mg/kg (via oral route; gastrointestinal effects). This substance may be absorbed by inhalation or ingestion; it irritates the skin and the respiratory tract; it is corrosive to the eyes. For prolonged exposures it exists a risk of cumulative effects.

Special precautions and preventive measures — Store in well-ventilated locations, away from any source of ignition (T<50°C). Wear rubber gloves and other protective laboratory clothes. Provide chemical respirators or a local exhaust system.

XCV. METHYL FLUOROACETATE

CAS RN 453189, CEE q: 1 kg.

Formula — $C_3H_5FO_2$.

Hazardous potentials — When heated to decomposition it emits toxic fumes of F.

Toxicity — It is highly toxic via oral, inhalation, subcutaneous, intramuscular, intraperitoneal, dermal, or parenteral routes.

LD_{50}: 4 mg/kg (administered to rat via subcutaneous routes).

LD_{50}: 300 mg/m³ (administered to rat via inhalation route over a period of 10 min).

LD_{50}: 5 mg/kg (administered to rat via subcutaneous route).

LD_{50}: 2500 μg/kg (administered to rat via intramuscular route).

LD_0: 5 mg/kg (administered to mouse via oral route).

LC_{50}: 3200 mg/kg (administered to mouse via inhalation route over a period of 10 min).

LD_{50}: 7500 μg/kg (administered to mouse via intraperitoneal route).

LD_{50}: 7 mg/kg (administered to mouse via subcutaneous route).

LD_{50}: 17 mg/kg (administered to mouse via intravenous route).

LD_{50}: 15 mg/kg (administered to mouse via parenteral route).

LDL_0: 100 μg/kg (administered to dog via oral route).

LC_{50}: 25 mg/m³ (administered to dog via inhalation route over a period of 10 min).

LDL_0: 100 μg/kg (administered to dog via subcutaneous route).

LD_{50}: 60 μg/kg (administered to dog via intravenous route).
LDL_0: 10 mg/kg (administered to monkey via oral route).
LCL_0: 800 mg/m³ (administered to monkey via inhalation route over a period of 10 min).
LDL_0: 10 mg/kg (administered to monkey via subcutaneous route).
LD_{50}: 4 mg/kg (administered to monkey via intravenous route).
LD_{50}: 300 μg/kg (administered to cat via oral route).
LD_{50}: 300 μg/kg (administered to cat via subcutaneous route).
LD_{50}: 500 μg/kg (administered to cat via intravenous route).
LC_{50}: 500 μg/kg (administered to rabbit via oral route).
LC_{50}: 100 mg/m³ (administered to rabbit via inhalation route over a period of 10 min).
LD_{50}: 20 mg/m³ (administered to rabbit by skin).
LDL_0: 300 μg/kg (administered to rabbit via subcutaneous route).
LD_{50}: 250 μg/kg (administered to rabbit via intravenous route).
LD_{50}: 400 μg/kg (administered to pig via intraperitoneal route).
LD_{50}: 500 μg/kg (administered to guinea pig via oral route).
LC_{50}: 100 mg/m³ (administered to guinea pig via inhalation route over a period of 10 min).
LD_{50}: 350 μ/kg (administered to guinea pig via intraperiotneal route).
LD_{50}: 200 μg/kg (administered to guinea pig via subcutaneous route).
LD_{50}: 2500 μg/kg (administered to hamster via intraperitoneal route).
LDL_0: 15 mg/kg (administered to chicken via intravenous route).

XCVI. METHYL ISOBUTYL KETONE PEROXIDE

CAS RN 37206205, CEE q: 50 t.
Formula — $C_{12}H_{26}O_6$.
Hazardous potentials — Severe explosion risk by impact, spontaneous chemical reactions, exposure to heat, or on contact with strong basis, metal salts, reducing, or flammable substances. As other peroxide it is a powerful oxidizer.
Toxicity — This substance irritates skin, eyes, and respiratory tract (mucous membrane).
Special precautions and preventive measures — Store in cool (T <30°C), well-ventilated locations, away from ignition sources or other substances. Wear chemical goggles, and other adequate protective clothes (shields, neoprene gloves, and aprons) necessary to prevent any skin or eye contact.

XCVII. METHYL ISOCYANATE

CAS RN 624839, CEE q:P 200 t. It is an intermediate product.
Flammability — Flammable. Flash point: less than −72°C (open cup); autoignition temperature: 535°C; and flammable limits in air, percent by volume: lower 5.3 and upper: 26. Dangerous when exposed to air, flame, or oxidizers.
Hazardous potentials — When heated to decomposition it emits toxic gases and vapors (such as hydrogen cyanide, oxides of nitrogen, and carbon monoxide). Contact with water causes formation of carbon dioxide and methylamine gases. The reaction is much more rapid in the presence of acids, alkalis, and amines. Contact with iron, tin, copper (or salts of these elements), and with certain other catalysts may cause violent polymerization.
Toxicity — Methyl isocyanate vapor is an intense lacrimator and irritates the eyes, mucous membranes, and skin. It can cause pulmonary irritation (edema) and sensitization. TLV-TWA: 0.02 ppm (skin) and OSHA standard air: TWA: 0.02 ppm = 0.05 mg/m³ averaged over an 8-h work shift. TCL_0: 2 ppm (irritant effects via inhalation route).
Special precautions and preventive measures — Methyl isocyanate will attack some

forms of plastics, rubber and coating. Wear impervious clothing, rubber gloves, face shields (8 in. minimum), and other appropriate protective clothing necessary to prevent any possibility of skin contact with liquid methyl isocyanate.

Disposal and waste treatment — "Spray into the furnace. Incineration will become easier by mixing with a more flammable solvent."[2]

XCVIII. *N*-METHYL-*N* 2,4,6-TETRANITROANILINE (TETRYL)

CAS RN 479458, CEE q: 50 t.
Formula — $C_7H_5N_5O_8$.
Flammability — As other nitrates it is dangerous when exposed to heat or flame.
Explosion — Severe explosion hazard when shocked or exposed to heat or flame. It is a powerful explosive, quite sensitive to percussion and more sensitive to shock and friction than TNT. It can be compressed into pellets for use as a booster explosive shells. Class A explosive. It reacts violently with hydrazine.
Hazardous potentials — On decomposition it emits toxic fumes of NO_x. As other nitrates it is a powerful oxidizing agent which may cause violent reaction with reducing materials.
Toxicity — It is irritant, sensitizer, and allergen. The chief effect produced by exposure to tetryl is the development of dermatitis. It may cause conjunctivitis, iridocyclitis, keratitis, or upper respiratory irritation. Tetryl workers may develop gastrointestinal symptoms, anemia, effects on the central nervous system, and hematopoietic and circulatory injuries. OSHA standard: air: TWA 1500 $\mu g/m^3$ and TLV-TWA: 1.5 mg/m^3 (skin). The time average TLV of 1.5 mg/m^3 for tetryl appears to be sufficiently low to prevent systemic poisoning but not sensitization.

XCIX. MEVINPHOS (α-2-CARBOMETHOXY-1-METHYL VINYL DIMETHYL-PHOSPHATE)

CAS RN 7786347, CEE q: 100 kg.
Formula — $C_7H_{13}O_6P$.
Hazardous potentials — When heated to decomposition it emits highly toxic fumes of PO_x. Severe explosion risk when heated.
Toxicity — TLV-STEL: 0.03 ppm (skin); TL-TWA: 0.01 ppm (skin); OSHA standard air: TWA 400 $\mu g/m^3$; and TDL_0 (oral-man): 690 $\mu g/kg/28$ D. Mevinphos is highly toxic via oral, inhalation, dermal, and intraperitoneal routes. It may affect the human peripheral nervous system. The greatest occupational hazard is absorption of the material through the skin, lungs, and mucous membranes. A TLV of 0.01 ppm should furnish a reasonable margin of safety, but the hazard of percutaneous absorption must be considered. The current suggested STEL of 0.03 mg/m^3 is under study for deletion. Acceptable daily intake 0.0015 mg/kg body weight.

Commodity	MRL (mg/kg)
Apples	0.5
Apricots	0.2
Beans	0.1
Broccoli	1
Brussel sprouts	1
Cabbage	1
Carrots	0.1
Cauliflower	1
Cherries	1
Citrus fruit	0.2
Collard	1

Commodity	MRL (mg/kg)
Cucumbers	0.2
Grapes	0.5
Lettuce	0.5
Melons	0.05
Onions	0.1
Peaches	0.5
Pears	0.2
Peas	0.1
Potatoes	0.1
Spinach	0.5
Strawberries	1
Tomatoes	0.2
Turnips	0.1

C. 4,4'-METHYLENE BIS-CHLOROANILINE (MOCA)

CAS RN 101144, CEE q: 10 kg.

Formula — $C_{13}H_{12}Cl_2N_2$.

Hazardous potentials — When heated to decomposition it emits very toxic fumes of Cl and NO_x.

Toxicity — OSHA: carcinogen and TLV-TWA: 0.02 ppm (skin). There is a possibility that high exposure may cause cancer in humans, most likely in the bladder or liver, therefore, 4,4'-methylene bis-(2-chloroaniline) must be considered an industrial substance suspected of carcinogenic potentials to humans. Moca exhibits the general toxicity characteristics of aromatic amines, cyanosis, methemoglobinemia, and kidney irritation. It is believed that the TLV of 0.02 ppm will prevent systemic poisoning, provided skin contact is avoided.

Preventive measures — Store in fire-resistive house; use adequate local exhaust ventilation and general ventilation.

Disposal and waste treatment — (1) Atomize in a suitable combusion chamber, (2) larger quantities can be disposed by diluting with water, then adding acetic acid in excess of that required to neutralize the propylene amine, or (3) pour into sodium bisulfate in a large evaporating dish. Sprinkle water and nuetralize. Drain into a sewer with sufficient water.

CI. α-NAPHTHYLAMINE

CAS RN 134327, CEE q.

Formula — $C_{10}H_9N$.

Flammability — Ignition temperature: 157°C and flash point: 315°F. Low fire hazard.

Explosion — Non-explosive.

Hazardous potentials — When heated to decomposition it emits toxic fumes of NO_x.

Toxicity — Naphthylamine is a human suspected carcinogen (IARC, OSHA). Aquatic toxicity rating TLm96: 10 to 1 ppm.

LD_{50}: 779 mg/kg (administed to rat via oral route).

LDL_0: 300 mg/kg (administered to rabbit via subcutaneous route).

Special precautions and preventive measures — Keep well closed and away from light. Wear respirator and protective clothes (rubber gloves).

Disposal and waste treatment — "(1) Dissolve in such combustible solvent as alcohols, benzene. Spray the solution into the furnace with after burner and scrubber, or (2) pour into a mixture of sand and soda ash. After mixing, put into a paper carton stuffed full with paper to serve as a fuel. Burn in the furnace."[2]

CII. β-NAPHTHYLAMINE (2-NAPHTHYLAMINE)

CAS RN 91598, CEE q: 1 kg.

Formula — $C_{10}H_9N$.

Flammability — Flash point 157°C. Combustible substance. Moderate fire hazard when exposed to heat or flame. At elevated temperature a vapor evolves, which is flammable or explosive.

Explosion — The explosive limits of this vapor have not yet been determined.

Hazardous potentials — When heated to decomposition it emits highly toxic fumes of NO_x.

Toxicity — It is noncorrosive or dangerously reactive but a very toxic chemical in any of its physical forms (flake, lump, dust, liquid, vapor). It can be absorbed through the lungs, the gastrointestinal tract, or the skin. Long and continued exposure may produce tumors and cancers of the bladder. It is a human carcinogenic substance via subcutaneous and oral routes. IARC carcinogenic determination: positive. OSHA: carcinogen agent; TLV: none (any contact with 2-naphthylamine should be avoided); and aquatic toxicity rating: TLm_{96}: 10 to 1 ppm.

TDL_0: 16 mg/kg (administered to rat via oral route over a period of 52 weeks intermittently; neoplastic effects).

TDL_0: 1300 mg/kg (administered to rat via intraperitoneal route, over a period of 13 weeks, intermittently; equivocal tumorigenic agent (ETA); this definition on the basis of studies reporting uncertain but seemingly positive results).

TDL_0: 7800 mg/kg (administered to rat via subcutaneous route over a period of 69 weeks intermittently; equivocal tumorigenic agent (ETA).

TDL_0: 20 mg/kg (administered to mouse via oral route over a period of 85 weeks; ETA).

TDL_0: 1600 mg/kg (administrated to mouse via subcutaneous route over a period of 40 weeks; ETA).

TDL_0: 18 mg/kg (administered to mouse via parenteral route; carcinogenic effects).

TDL_0: 100 mg/kg (administered to mouse by implant; ETA).

TDL_0: 18 mg/kg (administered to dog via oral route over a period of 2 years, intermittently; carcinogenic effects).

TDL_0: 17 mg/kg (administered to monkey via oral route; over a period of 5 years, intermittently; ETA).

TDL_0: 40 mg/kg (administrated to rabbit via oral route; over a period of 5 years, carcinogenic effects).

TDL_0: 217 mg/kg (administrated to hamster via oral route; over a period of 45 weeks; carcinogenic effects).

TD: 365 mg/kg (administrated to hamster via oral route; over a period of 43 weeks; carcinogenic effects).

TD: 105 mg/kg (administered to dog via oral route; over a period of 4 years; ETA).

Special precautions and preventive measures — Wear butyl rubber gloves, plastic working clothes, and self-contained breathing apparatus.

Disposal and waste treatment — See Section CI.

CIII. NICKEL CARBONYL

CAS RN 13463393, CEE q: 10 kg.

Formula — C_4NiO_4.

Flammability — Flammable; it is dangerous when exposed to heat, flame, oxidizers; it may cause fires and explosions. Flash point: lower than $-20°C$ (closed cup); autoignition temperature: may ignite spontaneously; and flammable limits in air: (percent by volume): lower 2 and upper: data not available.

Hazardous potentials — Liquid nickel carbonyl may explode when heated above 60°C. In the presence of air nickel carbonyl oxidizes and forms a deposit which becomes peroxidized. This tends to decompose and ignite. Contact with air, nitric acid, chlorine, O_2, Br_2, and other oxidizers may cause fires and explosions. The vapor of nickel carbonyl may promote the ignition of mixtures of combustible vapors (such as gasoline) and air. When heated (involved in a fire) or on contact with acid or acid fumes it emits highly toxic gases and vapors (nickel oxide, carbon monoxide).

Toxicity — Vapors may cause irritation, congestion, and edema of lungs. prolonged exposure may cause cancer of lungs and nasal sinuses. Highly toxic via inhalation and intraperitonal and intravenous routes. Sensitization and dermatitis is fairly common. Workers in a nickel refinery in which the Mond process is used, showed a high incidence of cancer of the lung and nasal sinus. TLV: 0.05 ppm and OSHA standard air: TWA: 7 μg/m^3 averaged over an 8-h work shift. NIOSH has recommended that nickel carbonyl be regulated as an occupational carcinogen. TCL_0: 7 mg/m^3 (central nervous system effects).

Special precautions and preventive measures — Nickel carbonyl may be formed inadvertently in situations where finely divided active nickel comes in contact with carbon monoxide. Liquid nickel will attack some forms of plastics, rubber, and coatings. Provide adequate ventilation with frequent checks of atmosphere for contamination. Store in a cool place away from sources of ignition. Wear rubber gloves, impervious, protective clothing (face shield), and a gas mask or airline respirator.

Disposal and waste treatment — Spray into a suitable combustion chamber.

CIV. NICKEL METAL

CAS RN 7440020, CEE q: 100 kg.

Formula — Ni.

Flammability — Nickel is a non flammable solid.

Hazardous potentials — Contact of nickel with strong acids may form flammable and explosive hydrogen gas. Contact with sulfur may cause evolution of heat. Contact of nickel nitrate with wood and other combustible may cause fire. Toxic gas and vapors (such as nickel carbonyl and oxides of nitrogen) may be released in a fire involving nickel or in the decomposition of nickel compounds.

Incompatibilities — Aluminum, aluminum trichloride, ethylene, *p*-dioxan, hydrogen, methanol, nonmetals, oxidants, and sulfur compounds. Reacts violently with F_2, NH_4NO_3, hydrazine, NH_3, P, and Se.

Toxicity — Metallic nickel or soluble nickel compounds may affect the body via inhalation, dermal, and oral routes. Metallic nickel and certain soluble compounds such as dust or fume cause sensitization dermatitis, allergic skin rashes, and probably produce cancer of the paranasal sinuses, of the lung; nickel fume in high concentration is a respirator irritant. Physical examinations of exposed personnel should be made available. TLV-TWA: 1.0 mg/m^3 Metal, 1.0 mg/m^3 insoluble compounds, as Ni, and 0.1 mg/m^3 soluble inorganic compounds. OSHA standard for nickel metal and soluble nickel compounds: 1 mg/m^3, averaged over an 8-h work shift. Occupational exposure to inorganic nickel recommended standard: air: TWA: 15 μ/m^3 averaged over a work shift of up to 10 h/d, 40 h/week. NIOSH has recommended that nickel be regulated as an occupational carcinogen.

Special precautions and preventive measures — Provide adequate ventilation. Wear thick gloves, impervious clothing, face shield, and other appropriate protective clothing necessary to prevent repeated or prolonged skin contact with powered metallic nickel or nickel compounds.

Disposal and waste treatment — ''Nickel metal and soluble nickel compounds may be disposed of in sealed containers in a secured sanitary landfill.''[3]

CV. NITROCELLULOSE

CAS RN 9004700, CEE q: 100 t.
Formula — $C_{12}H_{16}(ONO_2)O_6$.
Flammability — Flammable. Flash point: 55°F (13°C) and ignition temperature: 170°C. Extremely dangerous fire and explosion risk. Less flammable when wet.
Toxicity — Nitrocellulose has on human body effects similar to those that ethylic alcohol has.
Special precautions and preventive measures — Protect containers against physical damage. Drums should be protected against damage and not exposed to heating, nor should material be allowed to dry out. Storage should be segregated, well ventilated, and equipped with both decomposition and explosion vents, having the maximum amount of free opening. Protect against excessive heat and direct sunlight, avoid contact with electric light bulbs, steam coils, or other sources of heat. Wear rubber gloves, face shield, and self-contained breathing apparatus.
Disposal and waste treatment — ''Burn in a heat-resistant glass disk placed in a hood (25 ml maximum).''[2]

CVI. NITROGEN DIOXIDE

CAS RN 10102440, CEE q: 100 kg.
Formula — NO_2.
Flammability — Nitrogen dioxide is not combustible but it is strong oxidizing agent; it may cause fire in contact with clothing and other combustible materials.
Hazardous potentials — Toxic gases and vapors (such as oxides of nitrogen) may be released when nitrogen dioxide decomposes. Contact will all combustible materials, chlorinated hydrocarbons, ammonia, and carbon disulfide may cause fires and explosions.
Toxicity — Nitrogen dioxide gas is a respiratory irritant; chronic inhalation may cause headache, insomnia, ulcers of nose and mouth, anorexia, dyspepsia, dental erosion, weakness, chronic bronchitis, and emphysema. Nitrogen dioxide absorption may have effects on gastrointestinal, circulatory, and central nervous systems. TLV-TWA: 3 ppm; TLV-STEL: 5 ppm; and OSHA standard air: TWA 5 ppm. Occupational exposure to oxides of nitrogen recommended standard air: CL 1 ppm.
LCL_0: 200 ppm/1 min (via inhalation route).
TCL_0: 90 ppm/40 min (pulmonary effects when administered via inhalation route).
Special precautions and preventive measures — Protect containers against physical damage. Separate from combustible, organic, or other readily oxidizable materials. Since it is extremely corrosive when wet, proper materials for construction of the facilities are necessary.
Disposal and waste treatment — ''Spray or sift on a thick layer of a (1:1) mixture of dry soda ash and slaked lime behind a shield. After mixing, spray water from an atomizer with great precaution. Transfer slowly into a large amount of water. Neutralize and drain into the sewer with sufficient water.''[2]

CVII. NITROGEN MONOXIDE

CAS RN 10102439, CEE q: 100 kg.
Formula — NO.
Flammability — Nitrogen monoxide is not combustible, but it is strong oxidizing agent.
Hazardous potentials — It is dangerous; when heated to decomposition it emits highly toxic fumes of NO_x; it will react with water or steam to produce heat and corrosive fumes; can react vigorously with reducing materials.

Incompatibilities — Nitrogen monoxide can react violently (fire and explosions) with Al, B, CS_2, ClO, Cr, Fr, chlorinated hydrocarbons, Ammonia, O_3, P, and Na_2O.

Toxicity — It is highly irritant via inhalation and dermal routes and to eyes, and nervous membranes. TLV-TWA: 25 ppm averaged over a work shift of up to 10 h/d, 40 h/week. OSHA standard air: TWA 25 ppm (averaged over an 8-h work shift). Occupational exposure to oxides of nitrogen recommended standard air: CL 1 ppm averaged over 15 min period; TWA: 25 ppm.

Special precautions and preventive measures — Wear chemical goggles, rubber gloves and shields, and working protective clothing. Provide an adequate ventilation and self-contained breathing apparatus. The oxides of nitrogen are somewhat soluble in water, reacting with it in the presence of oxygen to form nitric and nitrous acids. This is the action that takes place deep in the respiratory tract.

CVIII. NITROGLYCERIN

CAS RN 55630, CEE q: 10 t.

Formula — $C_3H_5N_3O_3$.

Flammability — Dangerous when exposed to heat or flame or by spontaneous chemical reaction. Ignition temperature: 270°C.

Explosion — Explosion point 218°C. Severe explosion risk when shocked or exposed to O_3, heat, or flame. It is a powerful explosive very sensible to mechanical shock. Frozen nitroglycerine is somewhat less sensitive that the liquid. Class A Explosive.

Hazardous potentials — It emits toxic fumes on decomposition.

Toxicity — TLV-TWA: 0.05 ppm (skin); OSHA standard air: TWA: 2 mg/m^3 (skin); and occupational exposure to nitroglycerin: recommended. Standard air: CL 0.1 mg/mc for 20 min, 0.01 ppm to nitroglycerin and nitroglycol combined. If this material is taken internally, it causes respiratory difficulties and death due to respiratory paralysis. The most common complaint is headache which is noted commencing but soon passes off. Furthermore, nitroglycerin can be absorbed through uninjured skin and may produce eruptions. Toxic effects may occur by ingestion, inhalation of dust, or by absorption through intact skin. It may also cause vasodilation.

Special precautions and preventive measures — Store in standard combustible liquid storage. Keep away from initiator explosives. Separate from oxidizing materials, combustibles, and sources of heat. Wear neoprene gloves, plastic protective clothing, and self-contained breathing apparatus. Provide adequate ventilation.

CIX. NITROPHENYL DIETHYLPHOSPHATE

CAS RN 311455, CEE q: 100 kg

Formula — $C_{10}H_{14}NO_6P$.

Hazardous potentials — When heated to decomposition it emits highly toxic fumes of NO_x and PO_x.

Toxicity — In animals it is a highly toxic substance via oral, intraperitoneal, intravenous, and subcutaneous routes. It is a cholinesterase inhibitor (see Section CXII).

LD_{50}: 1800 μg/kg (administered to rat via oral route).

LD_{50}: 930 μg/kg (administered to rat via intraperitoneal route).

LD_{50}: 426 μg/kg (administered to mouse via subcutaneous route).

LD_{50}: 253 μg/kg (administered to rat via intravenous route).

LD_{50}: 446 μg/kg (administered to rat via intramuscular route).

LD_{50}: 19 mg/kg (administered to mouse via oral route).

LD_{50}: 330 μg/kg (administered to mouse via intraperitoneal route).

LD$_{50}$: 600 μg/kg (administered to mouse via subcutaneous route).
LD$_{50}$: 590 μg/kg (administered to mouse via intravenous route).
LD$_{50}$: 710 μg/kg (administered to mouse via intramuscular route).
LDL$_0$: 600 μg/kg (administered to mouse via parenteral route).
LDL$_0$: 100 μg/kg (administered to rabbit via intravenous route).
LDL$_0$: 300 μg/kg (administered to rabbit via subcutaneous route).
LDL$_0$: 2 mmg/kg (administered to chicken via oral route).

As other phosphorous compounds (except phosphine) it may cause serious disturbance, particularly in calcium metabolism.

CX. OXIDISULFOTON

CAS RN 2497076, CEE q: 100 kg.
Formula — $C_8H_{19}O_3PS_3$.
Hazardous potentials — When heated to decompostion it emits very toxic fumes of SO$_x$ and PO$_x$.
Toxicity — It is a cholinesterase inhibitor. It is highly toxic via oral route. It may be absorbed through skin.
LD$_{50}$: 3500 μg/kg (administered to rat via oral route).
LD$_{50}$: 193 mg/kg (administered to rat by skin).
LD$_{50}$: 12 mg/kg (administered to mouse via oral route).
LD$_{50}$: 263 mg/kg (administered to mouse by skin).

CXI. OXYGEN DIFLUORIDE (FLUORINE DIOXIDE (F_2O2), FLUORINE MONOXIDE (F_2O))

CAS RN 7783417, CEE q: 10 kg.
Formula — F_2O.
Flammability — Noncombustible.
Explosion — Oxygen difluoride is a powerful oxidizer. It explodes on contact with water, air, and reducing agents.
Hazardous potentials — When heated to decomposition it emits highly toxic fumes of fluorine.
Incompatibilities — On contact with all combustible materials, Cl_2, Br_2, I, NH_3, CO, I_2, H_2, Pt, and many other metals, metal oxides, and mixture with air it may cause fires and explosions.
Toxicity — Oxygen difluoride is very highly irritating to skin, eyes, and mucous membranes and corrosive to tissue. It attacks lungs with delayed appearance of symptoms. TLV: air: 0.05 ppm; OSHA standard air TWA: 100 μg/m^3 averaged over an 8-h work shift; and acceptable daily intake: 0.0005 mg/kg body weight. Maximum residue limits:

Commodity	MRL (mg/kg)
Almonds (on a shell-free basis)	0.1
Apples	2
Apricots	0.1
Beans	2
Cattle, carcass meat (in the carcass fat)	2.5
Cattle, edible offal (on a shell-free basis)	1
Cherries	0.1

Commodity	MRL (mg/kg)
Chestnuts (on a shell-free basis)	0.1
Citrus fruit	2
Cottonseed	0.5
Cucumbers	0.5
Eggplants (aubergines)	1
Eggs (on a shell-free basis)	0.2
Filberts (on a shell-free basis)	0.1
Garlic	1
Goats, carcass meat (in the carcass fat)	0.2
Goats, edible offal	0.2
Grapes	2
Horses, carcass meat (in the carcass fat)	0.2
Horses, edible offal	0.2
Maize (kernels)	0.05
Melons	2
Milk	0.02
Nectarines	1
Onions	1
Peaches	1
Pears	2
Pecans (on a shell-free basis)	0.1
Peppers	1
Pigs, carcass meat (in the carcass fat)	0.2
Pigs, edible offal	0.2
Pimentoes	1
Plums	2
Poultry (in the carcass fat)	
Poultry, edible offal	0.2
Sheep, carcass meat (in the carcass fat)	0.2
Sheep, edible offal	0.2
Squash	0.5
Strawberries	2
Tea (dry manufactured)	5
Tomatoes	2
Walnuts (on a shell-free basis)	0.1

Precautions and preventive measures — Oxygen difluoride will attack some forms of plastics, rubber, and coatings. Store away from reducing agents. Protect containers against physical damage. Long rubber gloves, goggles, protective clothing, and self-contained breathing apparatus may be used.

Disposal and waste treatment — "Spray or sift on a thick layer of a (1:1) mixture of dry soda ash and slaked lime behind a shield. After mixing, spray water from an atomizer with great precaution. Transfer slowly into a large amount of water. Neutralize and drain into the sewer with sufficient water."[2]

CXII. PARATHION

CAS RN 56382, CEE q: 100 kg.

Formula — $C_{10}H_{14}NO_5PS$.

Hazardous potentials — When heated to decomposition it emits highly toxic fumes of NO_x, PO_x, and SO_x. It reacts violently with endrin. Temperatures above 100°C may cause decomposition so that containers burst. Contact with strong oxidizers may cause fires and explosions. Parathion will attack some forms of plastics, rubber, and coatings.

Toxicity — Absorption may occur from inhalation of the vapor or mist, from skin absorption of the liquid, or from ingestion. Signs and symptoms of exposure are caused by the inactivation of the enzyme cholinesterase, which results in the accumulation of acetylcholine at synapses in the nervous system and skeletal and smooth muscle secretory glands. Parathion affects the central nervous system. Its effects are cumulative. Acceptable daily intake: 0.005 mg/kg body weight.

Commodity	MRL (mg/kg)
Apricots	1
Citrus fruit	1
Peaches	1
Other fruit	0.5
Vegetables	0.7
(except carrots)	

TLV-TWA: 0.1 mg/m³; OSHA standard air TWA: 110 μg/m³ (Poison); and occupational exposure to parathion recommended standard air TWA: 0.05 mg/m³.

LDL_0: 240 μg/kg (administered to man via oral route).

TDL_0: 5670 μg/kg (administered to woman via oral route).

LDL_0: 1471 μg/kg (administered to man).

Aquatic toxicity rating TLm^{96} under 1 ppm.

Special precautions and preventive measures — Store where the leakage from containers will not endanger the worker or contaminate other materials. Wear rubber gloves, self-contained breathing apparatus, and working clothing. Provide adequate ventilation. Weekly cholinesterase determinations of exposed workers are recommended.

Disposal and waste treatment — "Place the disposal in an open furnace. Add equal parts of sand and crushed limestone, then cover with a combustible solvent (alcohols, benzene, etc.). ignite from a safe distance and burn with the utmost care. (2) Place the disposal with sand and crushed limestone into a paper carton. Burn in the furnace with afterburner and alkali scrubber."[2]

CXIII. PARATHION METHYL (METHYL PARATHION)

CAS RN 298000.

Formula — $C_8H_{10}NO_5PS$.

Hazardous potentials — When heated to decomposition it emits very toxic fumes of NO_x, PO_x, and SO_x.

Toxicity — Parathion methyl is highly toxic via oral and inhalation routes. It may also be absorbed through skin. As other organophosphorous compounds it is a cholinesterase inhibitor. TLV-TWA: 0.2 mg/m³ (skin); occupational exposure to methyl parathion recommended standard: air TWA 0.2 mg/m³; and acceptable daily intake: 0.02 mg/kg body weight.

Commodity	MRL (mg/kg)
Cantaloupes	0.2
Cole crops	0.2
Cottonseed oil	0.05
Cucumbers	0.2
Fruit, other	0.2
Hops (dry cones)	0.05
Melons	0.2
Sugar beets	0.05
Tea (fermented and dried)	0.2
Tomatoes	0.2

Special precautions and preventive measures — Wear chemical goggles, respirator, rubber gloves and shoes, and other adequate protective clothing.

CXIV. PENTABORANE

CAS RN 19624227, CEE q: 100 kg.

Formula — B_5H_9.

Flammability — Flammable liquid by chemical reaction. It ignites spontaneously in air if impure. Flash point: 30°C (closed cup when pure); autoignition temperature: 35°C (impure material ignites spontaneously in air); and flammable limits in air: percent volume: lower: 0.42 and upper: not obtainable.

Explosion — Dangerous. Details unreported.

Hazardous potentials — On decomposition it emits toxic fumes and vapors (such as boron acids). Pentaborane reacts with oxidizers to form highly explosive mixture. Contact with halogens or halogenated compounds may cause explosions.

Incompatibilities — Dimethyl sulfoxide.

Toxicity — Highly toxic via inhalation and percutaneous routes. Pentaborane vapor affects the nervous system and causes signs of both hyperexcitability and narcosis. It can affect the body if it is inhaled, if it comes in contact with the eyes or skin, or if it is swallowed. TLV-TWA: 0.005 ppm; TLV-STEL: 0.015 ppm; and OSHA standard air: TWA 10 $\mu g/m^3$ averaged over an 8-h work shift.

Special precautions and preventive measures — Pentaborane will attack some forms of plastics, rubber, and coatings. Protect containers against physical damage. Store in well-ventilated, cool, dry location away from any area where the fire hazard may be acute. The storage tanks require a dry inert atmosphere of nitrogen at all times. Wear chemical goggles, clothing protected with rubber or Teflon, and self-contained breathing apparatus, face shields, and other appropriate protective clothing necessary to prevent skin contact with liquid Pentaborane, when skin contact may occur.

Disposal and waste treatment — "(1) Mix with dry sand. Transfer into a bucket and take into an open area. Spray dry butanol slowly. Spray water for complete destruction. Transfer into a large container. Neutralize with 8 *M*HCl. Decant the liquid and drain into a sewer with abundant water. Remove the residue of sand to a landfill. (2) Burn in open pit or iron pan with the utmost care."[2]

CXV. PENTAERYTHRITOL TETRANITRATE

CAS RN 78155, CEE q: 50 t.

Formula — $C_5H_8N_4O_{12}$.

Flammability — It is dangerous when exposed to heat or flame. Moderate fire hazard by spontaneous chemical reaction.

Explosion — Explosive Class A. Severe explosion hazard when shocked or exposed to heat. It is one of the most powerful high explosives, particularly sensitive to shock. It is used in detonating and priming compositions, as a base-charge in anticraft shells and mixed with TNT (70-30) in mines, explosive bombs, and torpedoes. It is a very effective demolition explosive.

Hazardous potentials — On decomposition it emits highly toxic fumes of NO_x. It can react vigorously with oxidizing materials.

Toxicity It produces skin effects if administrated via oral route. TLD_0: 1669 mg/kg (administered over a period of 8 years; skin effect).

CXVI. PEROXYACETIC ACID

CAS RN 79210, CEE q: 50 t.

Formula — CH_3COOOH.

Flammability — Combustible; moderate fire hazard via heat, flame, percussion. Flash point: 406°C ($\approx$105°F) (open cup).

Explosion — Severe explosion hazard when exposed to heat or by spontaneous chemical reaction. It is a powerful oxidizing agent. It may cause violent reactions with acetic anhydride, olefins organic matter. It explodes at 110°C.

Hazardous potentials — When heated to decomposition it emits acrid fumes.

Toxicity — Peroxyacetic acid is highly toxic and irritant. It is suspected to be carcinogenic by absorption through skin; it is an equivocal tumorigenic agent; this definition on the basis of studies reporting uncertain but seemingly positive results.

TDL_0: 21 mg/kg (administered to mouse by skin over a period of 26 weeks, intermittently in a number of discrete doses).

LD_{50}: 1540 mg/kg (administered to rat via oral route).

LD_{50}: 1410 mg/kg (administered to rabbit by skin).

LD_{50}: 10 mg/kg (administered to guinea pig via oral route).

Special precautions and preventive measures — Store in well-ventilated, cool, fire-resistant isolated room. Separate from other materials, especially from easily oxidizable organic materials and combustible materials. Avoid fire. For large quantity storage should be equipped with automatic sprinklers. Protect containers against physical damage. Wear rubber gloves, face shield, and coveralls.

Disposal and waste treatment — "Use vast volume of concentrated solution of reducing agent. Neutralize with soda ash or dilute HCl. Drain into a sewer with abundant water."[2]

CXVII. PHORATE

CAS RN 298022, CEE q: 100 kg.

Formula — $C_7H_{17}O_2PS_3$.

Hazardous potentials — Nonflammable. When heated to decomposition it emits highly toxic fumes of PO_x and SO_x.

Toxicity — Phorate is an organophosphorus cholinesterase inhibitor used as an insecticide. It is easily absorbed by all routes (oral, respiratory, and dermal). Phorate is a food additive permitted in the feed, in drinking water of animals, in treatment of food-producing animals. TLV-TWA: 0.05 mg/m³ (skin); TLV-STEL: 0.2 mg/m³ (skin); and LDL_0: 5 mg/m³ (via oral route).

Special precautions and preventive measures — Provide adequate ventilation. Wear chemical goggles, chemical respirator, rubber gloves and shoes, and other overall protective clothes.

CXVIII. PHOSGENE

CAS RN 75445, CEE q: 20 t.
Formula — $COCl_2$.
Hazardous potentials — Contact with moisture causes slow decomposition to form hydrogen chloride with carbon dioxide. Phosgene reacts violently with Al, K, Na, and Li; when heated to decomposition it emits toxic and corrosive fumes of chlorine and carbon monoxide, but these fumes are less toxic than phosgene itself.
Toxicity — Phosgene may affect the body via inhalation, dermal, or oral routes. It is a severe respiratory irritant. When it reaches the lung tissues it may cause pulmonary edema, which may be followed by bronchopneumonia and occasionally lung abscess. Concentration of 3 to 5 ppm of phosgene in air causes irritation of the eyes and throat, with coughing; 25 ppm is dangerous for exposure lasting 30 to 60 min, and 50 ppm is rapidly fatal after even short exposure. Skin contact with the liquid may cause severe burns. LC_{50} 3200 mg/min/m^3 (via inhalation route), and TCL_0 25 ppm/30 min (via inhalation route). Chemical warfare service: 1 ppm of phosgene may be considered safe for prolonged exposure.
Special precautions and preventive measures — Occupational exposure to phosgene may result not only from its use in the chemical industry, but as a result of the decomposition, in the presence of air or oxygen, of many chlorinated organic compounds. Store in a cool place, avoiding direct sunlight. Protect containers against physical damage. Wear long, rubber gloves, chemical goggles, chemical cartridge respirator, and a shield for whole body. Adequate ventilation, handle preferably in high-efficiency draft chambers or open outdoors surrounded by protective wall.

CXIX. PHOSPHINE (HYDROGEN PHOSPHINE)

CAS RN 7803512, CEE q: 100 kg.
Formula — H_3P.
Flammability — Autoignition temperature: 38°C. Phosphine is very dangerous by spontaneous reaction (in presence of P_2H_4); it ignites readily at room temperature when other hydrides of phosporus are present as impurities.
Explosion — Moderate explosion hazard when exposed to flame. It reacts violently with air, BCl_3, Br_2, Cl_2, HNO_3, HNO_2, and O_2. Contact with air or any other oxidizer (such as chlorine) may cause ignition of phosphine. It reacts with acids, halogenated hydrocarbons, and moisture.
Hazardous potentials — Dangerous when heated to decomposition it emits highly toxic fumes of PO_x; it can react vigorously with oxidizing materials.
Toxicity — It is highly toxic via inhalation route, but its action on the body has not fully worked out; it appears to cause chiefly a depression of the central nervous system and irritation of the lungs. TLV-STEL: 1 ppm; TLV-TWA: 3 ppm; OSHA standard air TWA: 400 μg/m^3; and LCL_0: 1000 ppm (administrated to man via inhalation route).
Special precautions and preventive measures — Provide adequate ventilation and adequate electric system. Store away from ignition sources or oxidizing materials. Wear chemical goggles, protective clothes (gloves, face shield), and a chemical cartridge respirator. Phosphine is an unusually reactive and poisonous substance, and is usually handled in small quantities only. A medical surveillance for exposed employees is recommended.

CXX. PHOSPHORAMIDOCYANIDIC ACID, DIMETHYL

CAS RN 63917419, CEE q: 1 t.
Formula — $C_3H_7N_2O_2P$.

CXXI. PHOSPHOTHIOIC ACID (*S*-(((1-CYANO-1-METHYL-ETHYL) CARBAMOYL)METHYL)-*O*,*O*-DIETHYLESTER)

CAS RN 3734950, CEE q: 100 kg.

Formula — $C_{10}H_{19}N_2O_4PS$.

Hazardous potentials — When heated to decomposition it emits highly toxic fumes of PO_x, SO_x, and NO_x.

Toxicity — It is highly toxic via oral and dermal routes. It is employed in agriculture as an insecticide.

LD_{50}: 3.5 mg/kg (administered to rat via oral route).

LD_{50}: 105 mg/kg (administered to rat by skin).

LD_{50}: 12 mg/kg (administered to mouse via oral route).

LD_{50}: 20 mg/kg (administered to dog via oral route).

LD_{50}: 8 mg/kg (administered to rabbit via oral route).

LD_{50}: 13 mg/kg (administered to guinea pig via oral route).

It is a cholinesterase inhibitor, like other organic phosphorous poisons.

Special precautions and preventive measures — Employees should be provided with respirator, and adequate protective clothing (chemical goggles, rubber gloves, and shoes) necessary to prevent any possibility of skin or eyes contact. Provide adequate ventilation.

CXXII. PICRAMIC ACID (SODIUM SALT)

CAS RN 831527, CEE q: 50 t.

Formula — $C_6H_4N_3O_5Na$.

Flammability — Flammable solid.

Hazardous potentials — When heated to decomposition it emits toxic fumes of No_x.

Toxicity — See Section CXXIII.

CXXIII. PICRIC ACID (2,4,6-TRINITROPHENOLO)

CAS RN 88891, CEE q: 50 t.

Formula — $C_6H_3N_3O_7$.

Flammability — Flash point: 150°C (closed cup); autoignition temperature: 300°C; and explosion temperature (required to cause explosion in 5 seconds): 322°C.

Hazardous potentials — Picric acid is shock sensitive; it forms highly sensitive explosives compounds (salts) with metals, particularly copper, lead, and zinc; these compounds are more sensitive to shocks than picric acid itself. Calcium salt, which is shock sensitive, may form when picric acid comes in contact with plaster or concrete. On decomposition it emits toxic gases and vapors, such as oxides of nitrogen and carbon monoxide. Picric acid is a more powerful explosive than TNT.

Toxicity — Picric acid may affect the body via inhalation, oral, or subcutaneous routes, or if it comes in contact with the eyes or skin. Exposure in industry is by skin contact or by inhalation of the dust of picric acid or its salts. The toxicology is of practical importance in the manufacture of munitions. The face is usually involved, especially around the mouth and sides of the nose. Toxic doses cause destruction of the erythrocytes and produce gastroenteritis, hemorrhagic nephritis, and acute hepatitis. The ingestion of 1 or 2 g in man causes severe poisoning. Picric acid dust causes sensitization, dermatitis, or allergic skin rashes. Preclude from exposure those individuals with diseases of liver, kidneys, and blood. Annual medical procedures should be made available to exposed personnel. TLV-TWA: 0.1 mg/m³ (skin); TLV-STEL: 0.3 mg/m³ (skin); and OSHA standard air TWA: 0.1 mg/m³ averaged over an 8-h work shift.

Special precautions and preventive measures — Protect from shock. Keep out of contact with metals. Employees should be provided with impervious clothing, butyl rubber gloves, face shield (8 in. minimum), protective clothing, and self-contained full breathing apparatus.

Disposal and waste treatment — "(1) Dissolve in such combustible solvent such as alcohols, benzene etc. Spray the solution into the furnace with after burner and scrubber. (2) Pour into sodium bicarbonate or a mixture of sand-soda ash (9:1). After mixing transfer into a paper carton filled with packing paper. Burn in an open furnace, or more efficiently in the furnace with afterburner and scrubber."[2]

CXXIV. PIRAZOXON (METHYL PYRAZOLYL DIETHYLPHOSPHATE)

CAS RN 108349, CEE q: 100 kg.
Formula — $C_8H_{15}N_2O_4P$.
Hazardous potentials — When heated to decomposition it emits very toxic fumes of NO_x and PO_x.
Toxicity — It is highly toxic via inhalation and oral routes and if it comes in contact with skin. It acts as cholinesterase inhibitor.
LD_{50}: 7 mg/kg (administered to rat via subcutaneous route).
LD_{50}: 4 mg/kg (administered to mouse via oral route).
LD_{50}: 40 mg/kg (administered to wild bird species via oral route).

CXXV. PROPANSULFONE (PROPANSULTONE; 1,2-OXATHIOLANE 2,2-DIOXIDE)

CAS RN 1120714, CEE q: 1 kg.
Formula — $C_3H_6O_3S$.
Hazardous potentials — When heated to decomposition it emits toxic fumes of SO_x.
Toxicity — Carcinogenic determination: animal positive (IARC). Propansultone is an industrial substance suspected of carcinogenic potential for man; no TLV has been assigned at this time, due to lack of epidemiological data on the effects of propansultone on workers.

CXXVI. PROPYLENE OXIDE (EPOXYPROPANE)

CAS RN 75569, CEE q: 50 t.
Formula — C_3H_6O.
Flammability — Propylene oxide is a flammable liquid, highly dangerous when exposed to heat or flame. Flash point: $-37°C$. Flammable limits: lower: 2.8% and upper: 37%.
Explosion — It is dangerous to explosion hazard, particularly when exposed to flame. Violent reaction with NH_4OH, clorosulfonic acid, HF, HNO_3, and NH_2SO_4.
Hazardous potentials — It polymerizes in contact with highly active catalysts with evolution of extensive heat, resulting in explosion when enclosed in vessel (chlorides or peroxides of iron, aluminum, hydroxides of alkalis). It can react vigorously with oxidizing materials.
Toxicity — Propylene oxide is classified toxicologically as a primary irritant, a mild protoplasmic protein, and a mild depressant of central nervous system activity. Contact with the skin, even with diluted propylene oxide may result in irritation and necrosis of the skin. Excessive exposure to the vapor irritates the eyes, upper respiratory tract, and lungs. It is used as an insecticidal fumigant agent and as a food additive permitted in food for human consumption. TLV-TWA: 20 ppm; OSHA standard: TWA 100 ppm; TLm_{96}: over 1000 ppm; and animal suspected carcinogen (IARC).
Special precautions and preventive measures — Use glass or metal container sealed with nitrogen. Detached outdoor storage is preferred. Store in a standard combustible liquid

storage room or cabinet. Separate from combustible materials or oxidizing materials. Provide adequate ventilation. Wear chemical goggles, large and heavy face shields (body shields are also recommendable), and self-contained breathing apparatus.

Disposal and waste treatment — ''(1) Place on ground in an open area. Evaporate or burn by igniting from a safe distance. (2) Dissolve in benzene, petroleum ether, or higher alcohol such as butanol. Dispose by burning the solution.''[2]

CXXVII. SELENIUM HEXAFLUORIDE

CAS RN 7783791, CEE q: 10 kg.
Formula — F_6Se.
Hazardous potentials — When heated to decomposition it emits very toxic fumes of F and Se.
Toxicity — Highly toxic via inhalation route. TLV-TWA: 0.05 ppm and OSHA standard air: TWA 400 $\mu g/m^3$. Long-term exposure can be a cause of amyotrophic lateral sclerosis in humans.
Special precautions and preventive measures — Process enclosure, personal protection equipment during the operations in which exposure may occur should be made available. Provide a local exhaust ventilation system.

CXXVIII. SODIUM CHLORATE

CAS RN 7775099, CEE q: 250 t.
Formula — ClO_3Na.
Flammability — Powerful oxidizing material. When involved in fire containers may be ruptured, followed by evolution of oxygen which supports burning, and of fumes (ClO_2) which are highly toxic.
Hazardous potentials — Explosive reactions with antimony sulfide, sulfur, sulfides, phosphorus, zinc, zinc chloride, and organic matter. Contact with concentrated H_2SO_4 or concentrated HNO_3 causes ignition or evolution of ClO_2. It also can react violently with static electricity, or paper.
Toxicity — Sodium chlorate damages the red blood corpuscles of women. Ingestion of large quantities can be fatal. It may cause local irritation to skin eyes and mucous membrane.
TDL_0: 800 mg/kg (administered to woman via oral route).
TLm_{96}: 1000 ppm.
LD_{50}: 185 mg/kg (administered to child).
Special precautions and preventive measures — Keep away from any source of ignition, and protect against heat, friction, percussion, and physical damage. Wear chemical goggles, rubber gloves, self-contained breathing apparatus, and a full protective shield.
Disposal and waste treatment — Use vast volume of concentrated solution of reducing agent. Neutralize with soda ash or dilute HCl. Drain into a sewer with abundant water.

CXXIX. SODIUM FLUORO ACETATE

CAS RN 62748, CEE q: 1 kg.
Formula — $C_2H_2FO_2Na$.
Flammability — Noncombustible.
Hazardous potentials — Toxic gases and vapors (such as hydrogen fluorides and carbon monoxides) may be released when sodium fluoroacetate decomposes.
Toxicity — Highly toxic via oral, dermal, inhalation, intraperitoneal, and subcutaneous routes. It is highly toxic both as a dust, or in water solution, it is used namely as a rodentricide.

It is slowly absorbed by skin unless the skin is abraided or cut, but rapidly absorbed by gastrointestinal tract. It operates by blacking the krebs cycle via formation of fluorocitric acid, which inhibits aconitase. It has an effect on either or both the cardiovascular and nervous systems in all species and in some species, the skeletal muscles. In man the action on the central nervous system produces epileptiform convulsive seizures followed by severe depression. The dangerous dose for man is 0.5v2 mg/kg. TLV-TWA: 0.05 mg/m^3 (skin); TLV-STEL: 0.15 mg/m^3 (skin); and OSHA standard air TWA: 50 μg/m^3 averaged over an 8-h work shift. LDL$_0$: 714 μg/kg (via oral route). The time-weighted average TLV of 0.05 mg/m^3 and the STEL of 0.15 mg/m^3 are low enough to prevent systemic toxicity.

Special precautions and preventive measures — Protect containers against physical damage. Wear rubber gloves, self-contained respirator, and full protective clothing (aprons, face shields, boots, chemical goggles or full face shield). It may cause explosions on shock with alkaline metals and carbon disulfide. Preclude from exposure individuals with diseases of central nervous system or heart.

Disposal and waste treatment — (1) Sodium fluoroacetate may be disposed of in a sealed container in a secured sanitary landfill. (2) Dissolve in a combustible solvent. Scatter spray of the solvent into furnace with afterburner and alkali scrubber.

CXXX. SODIUM SELENITE

CAS RN 10102188, CEE q: 100 kg.
Formula — O$_3$Sl-2Na.
Hazardous potentials — When heated to decomposition it emits toxic fumes of Se and Na$_2$O.
Toxicity — Sodium selenite is highly toxic via oral intraperitoneal, intravenous, subcutaneous, or intramuscular routes. Carcinogenic determination: indefinite (IARC). OSHA standard air TWA: 800 μg (Se)/m^3.
Special precautions and preventive measures — Avoid dust dispersion. Wear respirator, chemical goggles, gloves, and other adequate protective clothing. Store in dry locations.

CXXXI. SULFOTEP (ETHYL THIOPYROPHOSPHATE)

CAS RN 3689245, CEE q: 100 kg.
Formula — C$_8$H$_{20}$O$_5$P$_2$S$_2$.
Hazardous potentials — See Section CXII.
Toxicity — It is highly toxic via inhalation, oral, subcutaneous derman, intraperitoneal, intravenous, or intramuscolar routes. It is a cholinesterase inhibitor. Sulfotep is employed as a pesticide. TLV-TWA: 0.2 mg/m^3 (skin). OSHA standard air TWA 0.2 mg/m^3 (skin).

CXXXII. SULFUR DICHLORIDE

CAS RN 10545990, CEE q: 1 t.
Formula — Cl$_2$S.
Flammability — Moderate fire hazard when exposed to heat or flame. It has no flash point.
Hazardous potentials — It reacts violently with Al, NH$_3$, K, Na, oxidants, and metals. It is reactive with H$_2$O steam. When heated to decomposition it emits very toxic fumes of SO$_x$ and Cl.
Toxicity — Acquatic toxic rating TLm$_{96}$: 1000 to 100 ppm. OSHA standard air: 6 mg/m^3. It is highly irritant to skin eyes and mucous membrane and corrosive to tissue.
Special precautions and preventive measures — It is a corrosive material. Store in

places protected from moisture and water. Wear long rubber gloves, safety glasses, self-contained breathing apparatus, and working clothes.

Disposal and waste treatment — Spray on a thick layer of a (1:1) mixture of soda ash and slaked lime behind a shield. After mixing spray water from an atomizing with great precaution. Transfer slowly into a large amount of water. Neutralize and drain into a sewer with sufficient water.

CXXXIII. SULFUR DIOXIDE

CAS RN 7446095, CEE q: 1000 t.

Formula — SO_2.

Flammability — Not combustible.

Hazardous potentials — It will react with water or steam to produce toxic and corrosive fumes. Contact with some powdered metals and with alkaline metals such as sodium and potassium may cause fires and explosions.

Incompatibilities — Halogens, or interhalogens, lithium, nitrogen, metal acetylides, metal oxides, metals, polyumeric tubing, potassium chlorate, and sodium hydride.

Toxicity Sulfur dioxide is highly irritant via inhalation route and if it comes in contact with skin, eyes and mucous membrane. Its irritant properties are due to the rapidity with which, on contact with most membranes, it forms sulfurous acid very slowly oxidized to sulfuric acid. In combination with certain particulate matter and/or oxidants the effects may be markedly increased. High concentration may produce respiratory paralysis and pulmonary edema. Concentrations of 6 to 12 ppm cause immediate irritation of nose and throat; concentration of 400 to 550 ppm is immediately dangerous to life and 50 to 100 ppm is considered to be the maximum permissible concentration for exposures of 30 to 60 min. TLV-TWA: 2 ppm; OSHA standard air: TWA 5 ppm averaged over an 8-h work shift; and occupational exposure to sulfur dioxide, recommended standard: 0.5 ppm as a time weighted average for up to a 10-h work shift, 40-h work week.

LCL_0: 400 ppm/1 min (administered via inhalation route).

TCL_0: 4 ppm/1 min (administered to man via inhalation route; pulmonary effects).

Special precautions and preventive measures — Liquid sulfur dioxide will attack some forms of plastics, rubber, and coatings. Protect containers against physical damage, particularly from heat because elevated temperatures may cause them to burst. Outdoor storage or storage in fire-proofed, well-ventilated warehouses is preferable. Wear safety glasses, a dust mask, rubber gloves, and other appropriate protective clothing.

Disposal and waste treatment — ''(1) Pass the disposal gas through a large tank (drum, etc.) of soda ash solution. Add calcium hypochlorite with caution. Dilute and neutralize with 6 *M* HCl or 6 *M*NaOH. Drain into the sewer with abundant water. (2) Add the solid disposal to an equal volume of soda ash and dilute with water to obtain a slurry in a large container. Add calcium hypochlorite with caution. Dilute and neutralize with 6 *M* HCl or 6 *M* NaOH. Drain into the sewer with abundant water. (3) Seal cylinders and return to suppliers.''[2]

CXXXIV. TELLURIUM MEXAFLUORIDE

CAS RN 7783804, CEE q: 100 kg.

Formula — F_6Te.

Hazardous potentials — When heated to decomposition it emits very toxic fumes of F (hydrogen fluoride) and Te. It is dangerous when heated or on contact with acid or acid fumes.

Toxicity — Tellurium hexafluoride gas is a severe respiratory system and skin irritant.

No systemic effects have been reported from industrial exposure, but relatively few studies are available. TLV-TWA: 0.02 ppm; OSHA standard: TWA 0.2 mg/m$_3$ averaged over an 8-h work shift. TDL$_0$: 714 mg/kg (via inhalation route; skin effects).

Special precautions and preventive measures — Process enclosure. Provide an adequate local exhaust ventilation system, or a general dilution ventilation system. Respirators may be used for operations which require entrance into tanks or closed vessels or in emergency situations. Use protective equipment and clothing.

CXXXV. TEPP (TETRAETHYLPYROPHOSPHATE)

CAS RN 1070493, CEE q: 100 kg. Insecticide.
Formula — $C_8H_{20}O_7P_2$.
Flammability — Not combustible.
Hazardous potentials — When heated to decomposition it emits toxic fumes of PO_x, phosphoric acid mist, and carbon monoxide. Contact with strong oxidizers may cause fires and explosions.
Toxicity — The action results in an irreversible inhibition of the cholinesterase molecules and the consequent accumulation of large amounts of acetylcholine. Small doses at frequent intervals are largely additive. Absorption may occur from inhalation of the mist, by skin absorption, or by ingestion of the liquid. TLV-TWA: 0.05 ppm (skin) and OSHA standard TWA 50 μg/m^3 averaged over an 8-h work shift.
LDL$_0$: 2 mg/kg (via oral route).
TDL$_0$: 432 μg/kg (via oral route. Central nervous system effects).
LDL$_0$: 400 μg/kg (via intramuscular route).
TDL$_0$: 100 μg/kg CNS (via parenteral route). Central nervous system effects.
Special precautions and preventive measures — TEPP will attack some forms of plastics, rubber, and coatings. Provide an adequate ventilation. Weekly determinations of cholinesterase in exposed workers. Wear impervious clothing, rubber gloves, face shields (8 in. minimum), and other appropriate protective clothing necessary to prevent any possibility of skin contact with TEPP. Provide workers with self-contained breathing apparatus when necessary.
Disposal and waste treatment — (1) Place the disposal in an open furnace. Add to equal parts of sand and crushed limestone, then cover with a combustible solvent (alcohols, benzene, etc). Ignite from a safe distance and burn with the utmost care. (2) Place the disposal with sand and crushed limestone into a paper carton. Burn in the furnace with afterburner and alkaline scrubber. (3) Tepp may also be disposed by absorbing in vermiculite, dry sand, earth, or a similar material and disposing in scaled containers in a secured sanitary landfill.

CXXXVI. 2,3,7,8 TETRACHLORO-DIBENZO-*P*-DIOXIN (TCDD)

CAS RN 1746016, CEE q: 1 kg.
Formula — $C_{12}H_4Cl_4O_2$.
Toxicity — TLV: none established. Avoid any exposure. Carcinogenic determination: indefinite (IARC). It is a highly toxic substance via oral, dermal, and intraperitoneal routes. It causes death in rats by hepatic cell necrosis. Acute and subacute exposure may cause hepatic necrosis, thymic atrophy, hemmorhage lymphoid depletion, and chloracne.
Special precautions and preventive measures — Provide adequate ventilation. Wear chemical goggles, chemical cartridge respirator, and clean coveralls daily.

CXXXVII. TETRAETHYL LEAD

CAS RN 78002, CEE q: 50 t.
Formula — $C_8H_{20}Pb$.
Flammability — Flammable: moderate fire hazard where exposed to heat, flame, or oxidizers. Flash point 85°C (open cup) and 93°C (closed cup).
Explosion — Vapor forms explosive mixtures in air. Violently explodes when heated to more than 170°C.
Hazardous potentials — Tetraethyl lead can react vigorously with oxidizing materials, halogens. Hazardous decomposition products are triethyl lead (poisonous), carbon monoxide.
Toxicity — Tetraethyl lead can affect the body via inhalation, oral, or dermal routes and if it comes in contact with eyes. The fact that it is a lipoid solvent makes it an industrial hazard; it can cause intoxication not only by inhalation but also by absorption through the skin. Tetraethyl lead affects the nervous system and causes mental aberrations. It is a carcinogenic substance. TLV-TWA: 0.1 mg/m³ (skin); OSHA standard air TWA: 75 μg(Pb)/ m³ (skin); aquatic toxicity rating: TLm_{96} under 1 ppm; and carcinogenic determination: indefinite (IARC).
Special precautions and preventive measures — Storage should be in a cool, isolated, well-ventilated place. Outdoor or detached storage is preferred. Separate from halogenated compounds and oxidizing agents. Protect against all sources of ignition such as electric sparks and open flame. Provide adequate ventilation with regular monitoring of work areas. Wear chemical goggles, face shields, safety glasses, chemical cartridge or airline respirator, rubber gloves, and other protective, impervious clothing. Physical examinations should be made periodically available to exposed personnel.
Disposal and waste treatment — (1) Dissolve in a combustible solvent. Scatter the spray of the solution into the furnace with afterburner and alkali scrubber. (2) Tetraethyl lead may also be disposed by absorbing it in vermiculite, dry sand, earth, or a similar material and disposing in a secured sanitary landfill.

CXXXVIII. TETRAMETHYL LEAD

Formula — $Pb(CH_3)_4$.
Flammability — Tetramethyl lead is flammable; it is dangerous when exposed to heat, flame, or oxidizers. Flash point: 38°C.
Explosion — Moderate explosion hazard in the form of vapor when exposed to flame. It starts to decompose at 100°C and explodes violently at higher temperature (more stable than tetraethyl lead). Vapor may form explosive mixtures with air. Lower explosion limit: 1:8%.
Hazardous potentials — When heated it emits toxic gases and vapors (lead fumes, carbon monoxide). It can react vigorously with oxidizing materials (such as sulfuryl chloride or potassium permanganate).
Toxicity — Tetramethyl lead is highly toxic if it is inhaled, if it comes in contact with the eyes, or if it is swallowed. It may be absorbed through skin, but in this case it is moderately toxic. Tetramethyl lead is suspected to be carcinogen. It affects the central nervous system; intoxication resembles that caused by tetraethyl lead.
TLV-TWA: (in air) 0.15 mg/m³; (skin) and OSHA standard air TWA: 75 μg(Pb)/m³.
Special precautions and preventive measures — Protect containers against physical damage. Outdoor or detached and isolated storage is preferred. For inside storage combustible liquid storage rooms are preferred but separate from halides and other oxidizing materials. Vapor may form explosive mixtures with air. Protect against electric spark, open flame, or other ignition sources. Wear rubber gloves and self-contained breathing apparatus. Tetra-

methyl lead will attack some forms of rubber, plastics, and coatings. Physical examination should be made available to exposed personnel.

CXXXIX. TETRAMINE (TETRAMETHYLENEDISULFOTETRAMINE)

CAS RN 80126, CEE q: 1 kg.
Formula — $C_4H_8N_4O_4S_2$.
Hazardous potentials — When heated to decomposition it emits very toxic fumes of NO_x and SO_x.
Toxicity — It is highly toxic substance via oral and intraperitoneal routes.
LD_{50}: 20 µg/kg (administered to mouse via oral route).
LD_{50}: 4850 mg/kg (administered to rat via intraperitoneal route).
LD_{50}: 210 mg/kg (administered to mouse via intraperitoneal route).

CXL. TETRAZENE

CAS RN 109273, CEE q: 10 t.
Formula — $C_2H_8N_{10}O$.
Flammability — It is highly dangerous when exposed to heat or flame; moderate fire hazard by spontaneous chemical reaction.
Explosion — Severe explosion hazard when shocked or exposed to heat. It is a high explosive which evolves much flame. It is used in priming compositions and sometimes in combination with lead azide to lower the flash point of the azide.
Hazardous potentials — It is highly dangerous. When heated to decomposition it emits highly toxic fumes of NO_x and explodes.
Toxicity — As for other nitrates, large amounts taken by mouth may have serious or fatal effects. Small, repeated doses may lead to weakness, general depression, headache, and mental impairment. Also there is some implication of increased cancer incidence among those exposed.

CXLI. TRICHLOROMETHANE SULFENYL CHLORIDE

CAS RN 594423, CEE q: 100 kg.
Formula — CCl_4S.
Hazardous potentials — When heated to decomposition it emits highly toxic fumes of Cl and SO_x.
Toxicity — It is irritant to skin, eyes, and mucous membrane and via inhalation routes; it is an experimented cancerogen via inhalation route. TLV: air 0.1 ppm; OSHA standard air: TWA 800 µg/m³; and TCL_0: 45 ppm (via inhalation route; eye effects).

CXLII. TRINITROANISOLE (2,4,6 TRINITROANISOLO)

CAS RN 606359, CEE q: 50 t.
Formula — $CH_3N_3O_7$.
Flammability — As other nitrates dangerous when exposed to heat or flame.
Explosion — May explode by heat or shock.
Toxicity — It is considerably toxic. Symptoms: conjunctivitis, irritation of nose and throat, cephalalgia, fatigue, lack of appetite, and fever.
Special precautions and preventive measures — Wear butyl rubber gloves, protective working clothes, self-contained breathing apparatus, and protective shoes, and chemical goggles. Use adequate ventilation.

Disposal and waste treatment — "(1) Dissolve in such combustible solvent as alcohols, benzene, etc. Spray the solvent into the furnace with afterburner and scrubber. (2) Pour into sodium bicarbonate or a mixture of sand-soda (9:1). After mixing, transfer into a paper carton filled with packing paper. Burn in an open furnace, or more efficiently in the furnace with afterburner and scrubber."[2]

CXLIII. TRINITROBENZENE (1,3,5-TRINITROBENZENE)

CAS RN 25377326, CEE q: 50 t.
Formula — $C_6H_3N_3O_6$.
Flammability — Moderate fire hazard by spontaneous reaction.
Explosion — Class A explosive. Severe explosion hazard when shocked or exposed to heat; less sensitive than TNT to shock but has more shattering power than TNT. It is not used a lot because it is difficult to produce.
Hazardous potentials — It is a powerful oxidizing agent which may cause violent reactions with reducing materials. If decomposed by heating, poisonous nitrogen oxides will be evolved, which will violently react with reductive substance.
Toxicity — It is a highly toxic substance via oral route. Its principal effect, as for other nitrocompounds, results in reduction of the oxygen-carrying power of the blood, and depression of the central nervous system.
LD_{50}: 450 mg/kg (administered to rat via oral route).
LD_{50}: 730 mg/kg (administered to guinea pig via oral route).
LD_{50}: 505 mg/kg (administered to rat via oral route).
LD_{50}: 572 mg/kg (administered to mouse via oral route).
LD_{50}: 32 mg/kg (administered to mouse via intravenous routes).
Special precautions and preventive measures — It should be stored in a magazine, kept away from initiating explosives with surrounding kept vacant. Protect vessels against physical damage. Keep away from combustible materials, and sources of fire. Wear self-contained breathing apparatus.
Disposal and waste treatment — "(1) Dissolve in a combustible solvent. Spray the solution in a furnace with afterburner and scrubber. (2) Pour into sodium bicarbonate or a mixture of sand-soda ash. Burn in an open furnace with paper carton as combustible."[2]

CXLIV. 2,4,6-TRINITROBENZOIC ACID (DRY)

CAS RN 129668, CEE q: 50 t.
Formula — $C_7H_3N_3O_8$.
Hazardous potentials — When heated to decomposition it emits toxic fumes of NO_x. It exists as a hazard in preparation. Class A explosive.
Toxicity — As for other nitrates, large amounts taken by mouth may have serious or even fatal effects. The symptoms are dizziness, abdominal cramps, convulsions, and collapse. Small repeated doses may lead to weakness, general depression, headache, and mental impairment. Also there is some implication of increased cancer incidence among those exposed.

CXLV. 2,4,6 TRINITRO-*META*-CRESOL (CRESYLITE)

CAS RN 602993, CEE q: 50 t.
Formula — $C_7H_5N_3O_7$.
Flammability — As other nitrates it is dangerous when exposed to heat or flame.
Explosion — It exists a severe explosion hazard when it is shocked or exposed to heat.

Trinitrocresol is not as powerful a high explosive as TNT or picric acid. It has been used as a bursting charge and in combination with other high explosives.

Hazardous potentials — When heated to decomposition it emits highly toxic fumes of NO_x and it explodes. It can react vigorously with oxidizing materials.

Toxicity — It is highly toxic via intraperitoneal route. LDL_0: 31 mg/kg (administered to mouse via intraperitoneal route). LD_{50}: 168 mg/kg (administered to mouse via intraperitoneal route). As for other nitro compounds the principal effect results in reduction of the oxygen-carrying power of the blood and in depression of the central nervous system being responsible for most of the symptoms following acute exposure.

CXLVI. TRINITRORESORCINOLE (STYPHNIC ACID)

CAS RN 82713, CEE q: 50 t.
Formula — $C_6H_3N_3O_8$.
Explosion — Class A explosive.
Hazardous potentials — When heated to decomposition it emits toxic fumes of NO_x.
Toxicity — As for other nitro compounds the principal effect results in reduction of the oxygen-carrying power of the blood and in depression of the nervous system being responsible for most of the symptoms following acute exposure. Chronic poisoning occurs more frequently than acute; its principal effects are anemia, moderate cyanosis, insomnia, and loss of weight, complaints, related to nervous system.

CXLVII. sym-TRINITROTOLUENE

CAS RN 118967, CEE q: 50 t.
Formula — $C_7H_5N_3O_6$.
Flammability — As other nitrates it is dangerous where exposed to heat or flame. Wet trinitrotoluene is a flammable solid.

Explosion — Class A explosive. TNT will detonate under strong shocks. It detonates at around 240°C but can be distilled safely under reduced pressure. In small quantities it will burn quietly if not confined. However, sudden heating of any quantity will cause it to detonate; the accumulation of heat when large quantities are burning will cause detonation. In other respects it is one of the most stable of all high explosives and there are but few restrictions to its handling. It requires a fall of 130 cm for a 2-kg weight to detonate it. TNT containing an oxygen deficiency, the addition of products with are oxygen rich can enhance its explosive power. Exposure to light may increase sensitivity to shock.

Hazardous potentials — When heated to decomposition it emits highly toxic fumes of NO_x. It can react vigorously with reducing materials. Incompatibilities: sodium dichromate, sulfuric acid. Exposure to light, or contact with ammonia or with strong alkalies may increase sensitivity to shock. Contact with strong oxidizers may cause fire.

Toxicity — Trinitrotoluene is highly toxic via inhalation and oral routes; possible effects of exposure are dermatitis, cyanosis, gastritis, acute yellow atrophy of the liver, aplastic anemia, and occasionally, blood destruction, leukocytosis or leukopenia, varying degrees of central nervous system changes. It may be absorbed through skin, it is irritant to eyes and mucous membranes. TLV-TWA: 0.5 mg/m³ (skin) and OSHA standard air TWA: 1500 μg/m³ averaged over an 8-h work shift.

Special precautions and preventive measures — Store only in a permanent magazine. Keep away from initiator explosives. Protect containers against physical damage. Separate from oxidizing materials, combustibles, and sources of heat. Wear mechanical filter respirator, rubber gloves, and protective coveralls laundered daily. Periodic inspection of personnel for signs of cyanosis. Provide physical examinations every 6 months for exposed personnel.

Disposal and waste treatment — TNT may be disposed of only by explosive experts. (1) Dissolve in such combustible solvent as alcohols, benzene. Spray the solution into the furnace with afterburner and scrubber. (2) Pour in sodium bicarbonate or a mixture of sand-soda ash (9:1). After mixing, transfer into a paper carton. Burn in an open furnace, or more efficiently in the furnace with afterburner and scrubber.

CXLVIII. TRISAZIRIDINYLTRIAZINE

CAS RN 51183, CEE q: 10 kg.
Formula $C_8H_{12}N_6$.
Hazardous potentials — When heated to decomposition it emits highly toxic fumes of NO_x.
Toxicity — It is a highly toxic substance via oral, intraperitoneal, intravenous, or intramuscular routes. It has been reported as capable of causing gastrointestinal disturbances and bone marrow depression. Carcinogenic determination: animal positive (IARC).

TDL_0: 550 μg/kg (administered via intraperitoneal route to rat over a period of 12 d).

TDL_0: 1 mg/kg (administered to rat via intraperitoneal route over a period of 5 d).

TDL_0: 1350 μg/kg (administered to mouse via subcutaneous route over a period of 7 to 9 d; teratogenic effects).

TDL_0: 2800 μg/kg (administered to mouse via oral route).

TDL_0: 1 mg/kg (administered to mouse via intraperitoneal route).

TDL_0: 300 mg/kg (administered to mouse via intraperitoneal route).

TDL_0: 200 μg/kg (administered to rabbit via intravenous route).

TDL_0: 500 μg/kg (administered to rabbit via intraperitoneal route).

TDL_0: 500 μg/kg (administered to rat via intraperitoneal route; teratogenic effects).

TDL_0: 10 mg/kg (administered to rat via subcutaneous route; equivocal tumorigenic agent; this definition on the basis of studies reporting uncertain but seemingly positive results).

TDL_0: 10 mg/kg (administered to mouse by skin; neoplastic effects).

TDL_0: 3 mg/kg (administered to rat via intraperitoneal route, over a period of 19 d, intermittently; neoplastic effects).

TD: 6 mg/kg (administered to mouse via intraperitoneal route, over a period of 42 d, intermittently; neoplastic effects.

LD_{50}: 1 mg/kg (administered to rat via oral route).

LD_{50}: 1 mg/kg (administered to rat intraperitoneal route).

LD_{50}: 1110 μg/kg (administered to rat via intravenous route).

LD_{50}: 1500 μg/kg (administered to rat via intramuscular route).

LD_{50}: 15 mg/kg (administered to mouse via oral route).

LD_{50}: 2800 μg/kg (administered to mouse via intraperitoneal route).

LD_{50}: 1500 μg/kg (administered to mouse via intramuscular route).

LDL_0: 1 mg/kg (administered to dog via oral route).

LDL_0: 400 μg/kg (administered to dog via intravenous route).

LDL_0: 100 μg/kg (administered to monkey via intravenous route).

LDL_0: 1 mg/kg (administered to cat via intravenous route).

CXLIX. WARFARIN (COUMADIN)

CAS RN 81812, CEE q: 100 kg.
Formula — $C_{19}H_{16}O_4$.
Toxicity — Warfarin has two actions: inhibition of prothrombin formation and capillary damage. It is highly toxic via inhalation route, and moderately toxic via oral and dermal routes. Warfarin has cumulative effects. TDL_0: 26 mg/kg (administered to woman via oral route; teratogenic effects). TLV-TWA: 0.1 mg/m³; TLV-STEL: 0.3 mg/m³; and OSHA standard air: TWA 100 μg/m³.

Index

INDEX

Milton Keynes UK
Ingram Content Group UK Ltd.
UKHW052017071024
449327UK00027B/2317